# THE ETHICS OF FOOD

A Reader for the Twenty-First Century

EDITED BY
GREGORY E. PENCE

ROWMAN & LITTLEFIELD PUBLISHERS, INC.
*Lanham • Boulder • New York • Oxford*

ROWMAN & LITTLEFIELD PUBLISHERS, INC.

Published in the United States of America
by Rowman & Littlefield Publishers, Inc.
4720 Boston Way, Lanham, Maryland 20706
www.rowmanlittlefield.com

12 Hid's Copse Road
Cumnor Hill, Oxford OX2 9JJ, England

Copyright © 2002 by Rowman & Littlefield Publishers, Inc.

*All rights reserved.* No part of this publication may be reproduced, stored in a retrieval system, or transmitted in any form or by any means, electronic, mechanical, photocopying, recording, or otherwise, without the prior permission of the publisher.

British Library Cataloguing in Publication Information Available

**Library of Congress Cataloging-in-Publication Data**

The ethics of food : a reader for the twenty-first century / edited by Gregory E. Pence.
    p. cm.
    Includes bibliographical references and index.
    ISBN 0-7425-1333-5 (alk. paper)—ISBN 0-7425-1334-3 (pbk. : alk. paper)
    1. Genetically modified foods—Moral and ethical aspects.  2. Nutrition policy—Moral and ethical aspects.  I. Pence, Gregory E.

TP248.65.F66 E86   2002
363.19'2—dc21                                                                                          2001048271

Printed in the United States of America

∞ ™ The paper used in this publication meets the minimum requirements of American National Standard for Information Sciences—Permanence of Paper for Printed Library Materials, ANSI/NISO Z39.48-1992.

# CONTENTS

Introduction: The Meaning and Ethics of Food   GREGORY E. PENCE   vii

## Part I: The Meaning of Food

CHAPTER 1

A Thing Shared   M. F. K. FISHER   1

CHAPTER 2

How We Grow Food Reflects Our Virtues and Vices   WENDELL BERRY   5

## Part II: Eating Meat

CHAPTER 3

Animal Liberation and Vegetarianism   PETER SINGER   26

CHAPTER 4

Meat Is Good for You   STUART PATTON   51

## Part III: Starvation

CHAPTER 5

Lifeboat Ethics: The Case against Helping the Poor   GARRETT HARDIN   54

CHAPTER 6

Golden Rice Is Fool's Gold   GREENPEACE INTERNATIONAL   71

CHAPTER 7

Are We Going Mad?   NORMAN BORLAUG   74

## Part IV: Safety of Genetically Modified Foods

CHAPTER 8

The Unholy Alliance   MAE-WAN HO   80

CHAPTER 9

The FDA's Volte-Face on Food Biotech   HENRY I. MILLER   96

CHAPTER 10

Dr. Strangelunch: Why We Should Learn to Love Genetically Modified Food   RONALD BAILEY   100

## Part V: Benefits/Dangers of Organic Food

CHAPTER 11

Organic or Genetically Modified Food: Which Is Better?   GREGORY E. PENCE   116

CHAPTER 12

The Benefits of Organic Food   TANYA MAXTED-FROST   123

## Part VI: Genetically Modified Food and Environmental Risks

CHAPTER 13

Genetic Engineering and Food Security   VANDANA SHIVA   130

CHAPTER 14

GM Food Is the Best Option We Have   ANTHONY J. TREWAVAS   148

## Part VII: Food Biotechnology and Nature

CHAPTER 15

Biotechnology's Negative Impact on World Agriculture   MARC LAPPÉ AND BRITT BAILEY   156

CHAPTER 16

The Population/Diversity Paradox: Agricultural Efficiency to Save Wilderness   ANTHONY J. TREWAVAS   168

## Part VIII: Global Food Politics and Economics

CHAPTER 17

A Removable Feast   C. FORD RUNGE AND BENJAMIN SENAUER   180

CHAPTER 18

From Global to Local: Sowing the Seeds of Community   HELEN NORBERG-HODGE, PETER GOERING, AND JOHN PAGE   191

## Part IX: The Food Industry

CHAPTER 19

The Hamburger Bacteria   NICOLS FOX   215

CHAPTER 20

The United States Food Safety System   U.S. FOOD AND DRUG ADMINISTRATION   267

Index   281

About the Editor   287

# Introduction
# The Meaning and Ethics of Food

GREGORY E. PENCE

FOOD MAKES PHILOSOPHERS of us all. Death does the same, but most of us try to avoid thinking about death. Of course death comes only once, so we can postpone thinking about it, but choices about food come many times a day, every day.

Like sex, food is an essential aspect of human experience. Like sex, the decisions we make about food define who we have been, who we are now, and who we want to become. How we make those choices says much about our values, our relationship to those who produced our food, and the kind of world we want.

For much of human history, just getting enough food was a struggle, one often lost. After getting enough to eat, humans wanted to store any excess food for future bad times. Reliable, secure access to food is a relatively new phenomenon in civilization, as the first settlers in Jamestown, Virginia, must have realized as they slowly died of starvation. In the nineteenth century "vital amines," or vitamins, were discovered and, as such, became the foundations of modern nutrition. People began to realize that what made you healthy was not just how much food you ate or how often but what kind.

Today, information about food abounds and, as a result, we face a plethora of choices every day: Do we eat a fast-food hamburger or a soyburger with bean sprouts? Do we eat fresh sushi and try our luck on new ethnic restaurants, such as one offering Indonesian fare? Do we buy locally produced produce, cheese, and meat or buy globally, perhaps using the Internet to order? Do we eat meat at all? Processed food? Sugar? Sweet 'N Low?

Given all these choices, it is remarkable that philosophers, bioethicists, sociologists, and others have not written more about food. Food is life. To continue living, all life must consume some kind of food. Upon this brute fact, humans over thousands of years have heaped countless layers of meaning.

In the main, writing about food has been left to freelance writers or those who write about food as an economic commodity. As this collection hopes to show, there is much more to write about food than its status as an item of agricultural exchange. Indeed, we can think about food at many different levels: the molecular and chemical, the medical, the personal, the nonmoral evaluative, the moral, the economic, the political, and the global. Each level has its own issues interwoven with those of the other levels.

In this book, little will be said of the viewpoint of physics where food is just the exchange of energy, or of biochemistry, where eating food is partly a chemical reaction mediated by enzymes and proteins. Of course the personal level is the one most familiar to each of us, and decisions about eating desserts, meat, fast food, and organic foods figure powerfully in our lives.

Because food is so necessary to life, the conveyance of food from parents to children has primordial significance. The most powerful symbol of this significance is breast-feeding, in which the mother literally gives of herself to sustain the infant. Conveyed in this action are altruism, pleasure, nurturing, love, and safety. The famous writer M. F. K. Fisher conveys such meanings of food in her short piece, "A Thing Shared," which begins this volume. Following this piece is a selection from American essayist Wendell Berry, who is also a Kentucky farmer and poet. Berry extols the bygone virtues of American farm life and bemoans the vices of modern agribusiness and industrial foods.

Food must be transformed before it is consumed. Even primitive man quickly learned not to immediately eat raw food but to give thanks to the gods, probably by offering the first piece of meat to the fire. In Christianity, food must be blessed before it is consumed or, at least, a blessing must be said before the family eats. In Orthodox Judaism, kosher meat must come from animals killed under the supervision of a rabbi. In Islam, pork is forbidden; in Hinduism, cows cannot be killed.

But should we eat "animal flesh" at all? In "Animal Liberation and Vegetarianism," Princeton bioethicist Peter Singer argues that for moral reasons we should boycott eating meat to protest the suffering of animals raised for human consumption in our industrial food system. In "Meat Is Good for You," food scientist Stuart Patton argues that there is nothing wrong with eating meat and that it is even healthy for us to eat it.

Perhaps the most important questions about food today are the various issues raised by genetically modified (GM) food. The phrase "genetically modified food" raises two dangers at once: it is "genetic" and, hence, is feared to cause cancer in the same way that human genes can cause cancer, while food is "environmental" in being something that is ingested and that acts on one's existing body. So GM food is feared on both counts.

Tragedy in ethics occurs when we are faced with two decisions where great evil occurs either way we choose. Sometimes fate throws such tragedies our way, as Oepidus learned, or in the *Godfather* saga, when Michael Corleone orders his brother Fredo to be killed. When we think of starving people and GM food, some critics say we are in such a tragic situation: either we do not create GM foods and let millions starve, or we do create GM foods and risk ruining the environment.

Such issues are discussed by biologist Garrett Hardin, who believes that feeding starving people now (with GM foods or other inventions) will only delay starvation and create more starving people later. Greenpeace International also opposes using genetically modified crops to help feed the starving. It has led the worldwide campaigns to keep GM foods and crops out of England, Europe, and Africa, and now advocates the same for North America. In contrast, Nobel Prize–winning biologist Norman Borlaug argues that in this century, scientific agriculture has conquered starvation and can do so again with GM foods.

On a more global level, Indian activist and scientist Vandana Shiva aligns with Greenpeace to attack the safety and environmental benefits of GM food as well as the mindset of Western science and international biotechnology corporations. Great Britain and Europe have closed the door on GM foods and GM crops. Are they wise or foolish? London-based biologist Mae-Wan Ho argues that GM foods are totally unacceptable because they are unhealthy and bad for the environment, they exploit indigenous peoples, and they reduce something holy to a commodity. Former U.S. Food and Drug Administration commissioner Henry I. Miller disagrees, arguing that GM foods have been better tested than any other foods in history and that they are as safe as any conventional foods. Ronald Bailey, a science writer and editor at *Reason*, argues that critics of GM food are not really opposed to this food on grounds of safety but are really attacking international corporations, capitalism, and intellectual property rights in biotechnology.

Bioethicist and volume editor Gregory Pence defends GM food against some of the common attacks against it by contrasting its alleged dangers against the assumed safety of organic food. Pence argues that GM food is actually *safer* than organic food. Tanya Maxted-Frost's paean to organic food espouses the opposite view and is typical of writings about organic food.

Another fear about GM plants is that they will ruin the environment.

Such fears are certainly raised by reading the work of Vandana Shiva, an Indian physicist turned food critic who wants the world's agriculture to return to the pure, organic ways of indigenous peoples, thereby protecting biodiversity and local control. In this desire, she aligns herself with Greenpeace International, Mae-Wan Ho, and Wendell Berry.

Against their viewpoints, Scottish professor Anthony Trewavas, a plant physiologist, argues that the environment can actually be made safer by using GM crops. In pushing this theme, he echoes a theme of Norman Borlaug's, who argues that a worldwide expansion of organic farming would destroy millions of extra acres of the world's forest and prairie.

But there are more values at stake here than simply feeding people versus making the environment safer. What kind of ecosystem do we want? How valuable is biodiversity? "Wildness?" How much "wildness" do we want to preserve and at what cost? Alternately, are huge monocultures of coffee in Colombia, sugar in Cuba, pineapples in Hawaii, and shrimp off Ecuador's coast a bad thing? Americans Marc Lappé, a geneticist and founder of an environmental think tank in California, and his associate Britt Bailey argue that GM foods and agricultural biotechnology are likely to turn out badly for the environment, for people, and for farming. They use historical examples to make their case. In contrast, Anthony Trewavas argues that biotechnology is not opposed to food safety, preserving the environment, or helping farmers but, in fact, is the best tool we have to do all of these.

Economic globalists and libertarians believe that a rising sea lifts all boats, so the best way to preserve the environment, grow a democracy, and feed people is to grow a healthy economy with a strong middle class. Seen from this view, GM food can be a tool toward helping indigenous peoples move toward self-sustaining, democratic economies. So argue C. Ford Runge and Benjamin Senauer in this volume. Not everyone agrees with them, especially Helen Norberg-Hodge, Peter Goering, and John Page from the International Society for Ecology and Culture, who argue against globalism, capitalism, and the new biotechnology owned by international agribusiness.

With the germ theory of disease at the beginning of the twentieth century, a new kind of purification ritual began. Now food must be purified of germs and scientifically sanitized. Periodic outbreaks of *E. coli* in hamburger meat, such as the one that occurred at fast-food chains in Seattle in 1998 or the one in large meat-processing plants in 1999, remind us that food can spoil and be infected by pathogens. The selection from Nichols Fox in this volume discusses the special dangers of contaminated food in our vast industrial food system. Opposed to Fox is the selection from the Food and Drug Administration, which represents the official position of the American government and food industry that food in America is the safest in the world.

Taken together, the twenty pieces in this collection, grouped around nine general topics, represent a fascinating, wide-ranging view of modern food and its meaning to humans. It is the editor's hope that this collection will stimulate new courses on the ethics of food or, at least, incorporation of this topic into traditional courses that discuss applied ethics and bioethics. It is also the editor's hope that the intelligent reader will find this collection of lasting value. How we think about food really is important, and such thinking helps define who we are and who we want to become, both as individuals and as a common humanity.

# A Thing Shared

M. F. K. FISHER

*As Clifton Fadiman explained in 1954 in his introduction to Fisher's justly famous* The Art of Eating *(a collection of five books of her essays on food), M. F. K. Fisher is "the most interesting philosopher of food now practicing in this country.... Here then is a witty, well-furnished mind roving over the field of food, relating it to the larger human experience of which dining, while a miracle, is still but a part." Fadiman's judgment was echoed by W. H. Auden, who in 1963 called her "America's greatest writer."*

*Fisher's essays show how eating is not just consuming calories but an act that defines who we are and how we see the world and its inhabitants. Dismissed by many for years as "a woman's food writer," Fisher had a real message that was about food, cooking, and eating as human and cultural metaphors.*

*Her writing career spanned sixty years and many of her essays first appeared in the* New Yorker *magazine. She published fifteen books during this career, many of them collections of essays. In a review of one of her books in 1982, Raymond Sokolov wrote, "In a properly run culture, Mary Frances Kennedy Fisher would be recognized as one of the great writers this country has produced in this century." Fisher died on her California ranch in 1992 at the age of eighty-three.*

*In "A Thing Shared," Fisher writes about the first time she and her sister, as young girls, ate a meal alone with their father without their supervising mother. This meal occurred on a hot August day as a spontaneous picnic along a freshwater stream in a rural California canyon. Twenty-five years later, all three adults would remember the meal both as one of the best of their lives and as one that changed them.*

NOW YOU CAN DRIVE from Los Angeles to my Great-Aunt Maggie's ranch on the other side of the mountains in a couple of hours or so, but the first time I went there it took most of a day.

From *The Art of Eating* by M. F. K. Fisher. Copyright © 1990 by M. F. K. Fisher. All rights reserved. Reproduced here by permission of the publisher.

Now the roads are worthy of even the All-Year-Round Club's boasts, but twenty-five years ago, in the September before people thought peace had come again, you could hardly call them roads at all. Down near the city they were oiled, all right, but as you went farther into the hills toward the wild desert around Palmdale, they turned into rough dirt. Finally they were two wheel-marks skittering every which way through the Joshua trees.

It was very exciting: the first time my little round brown sister Anne and I had ever been away from home. Father drove us up from home with Mother in the Ford, so that she could help some cousins can fruit.

We carried beer for the parents (it exploded in the heat), and water for the car and Anne and me. We had four blowouts, but that was lucky, Father said as he patched the tires philosophically in the hot sun; he'd expected twice as many on such a long hard trip.

The ranch was wonderful, with wartime crews of old men and loud-voiced boys picking the peaches and early pears all day, and singing and rowing at night in the bunkhouses. We couldn't go near them or near the pen in the middle of a green alfalfa field where a new prize bull, black as thunder, pawed at the pale sand.

We spent most of our time in a stream under the cottonwoods, or with Old Mary the cook, watching her make butter in a great churn between her mountainous knees. She slapped it into pats, and put them down in the stream where it ran hurriedly through the darkness of the butter-house.

She put stone jars of cream there, too, and wire baskets of eggs and lettuces, and when she drew them up, like netted fish, she would shake the cold water onto us and laugh almost as much as we did.

Then Father had to go back to work. It was decided that Mother would stay at the ranch and help put up more fruit, and Anne and I would go home with him. That was as exciting as leaving it had been, to be alone with Father for the first time.

He says now that he was scared daft at the thought of it, even though our grandmother was at home as always to watch over us. He says he actually shook as he drove away from the ranch, with us like two suddenly strange small monsters on the hot seat beside him.

Probably he made small talk. I don't remember. And he didn't drink any beer, sensing that it would be improper before two unchaperoned young ladies.

We were out of the desert and into deep winding canyons before the sun went down. The road was a little smoother, following streambeds under the live-oaks that grow in all the gentle creases of the dry tawny hills of that part of California. We came to a shack where there was water for sale, and a table under the dark wide trees.

Father told me to take Anne down the dry streambed a little way. That made me feel delightfully grown-up. When we came back we held our hands under the water faucet and dried them on our panties, which Mother would never have let us do.

Then we sat on a rough bench at the table, the three of us in the deep green twilight, and had one of the nicest suppers I have ever eaten.

The strange thing about it is that all three of us have told other people that same thing, without ever talking of it among ourselves until lately. Father says that all his nervousness went away, and he saw us for the first time as two little brown humans who were fun. Anne and I both felt a subtle excitement at being alone for the first time with the only man in the world we loved.

(We loved Mother too, completely, but we were finding out, as Father was too, that it is good for parents and for children to be alone now and then with one another . . . the man alone or the woman, to sound new notes in the mysterious music of parenthood and childhood.)

That night I not only saw my Father for the first time as a person. I saw the golden hills and the live-oaks as clearly as I have ever seen them since; and I saw the dimples in my little sister's fat hands in a way that still moves me because of that first time; and I saw food as something beautiful to be shared with people instead of as a thrice-daily necessity.

I forget what we ate, except for the end of the meal. It was a big round peach pie, still warm from Old Mary's oven and the ride over the desert. It was deep, with lots of juice, and bursting with ripe peaches picked that noon. Royal Albertas, Father said they were. The crust was the most perfect I have ever tasted, except perhaps once upstairs at Simpson's in London, on a hot plum tart.

And there was a quart Mason jar, the old-fashioned bluish kind like Mexican glass, full of cream. It was still cold, probably because we all knew the stream it had lain in, Old Mary's stream.

Father cut the pie in three pieces and put them on white soup plates in front of us, and then spooned out the thick cream. We ate with spoons too, blissful after the forks we were learning to use with Mother.

And we ate the whole pie, and all the cream . . . we can't remember if we gave any to the shadowy old man who sold water . . . and then drove on sleepily toward Los Angeles, and none of us said anything about it for many years, but it was one of the best meals we ever ate.

Perhaps that is because it was the first conscious one, for me at least; but the fact that we remember it with such queer clarity must mean that it had other reasons for being important. I suppose that happens at least once to every human. I hope so.

Now the hills are cut through with super-highways, and I can't say whether we sat that night in Mint Canyon or Bouquet, and the three of us are in some ways even more than twenty-five years older than we were then. And still the warm round peach pie and the cool yellow cream we ate together that August night live in our hearts' palates, succulent, secret, delicious.

# How We Grow Food Reflects Our Virtues and Vices

## 2

WENDELL BERRY

*Wendell Berry, born in 1934, grew up on a working farm in Kentucky and since 1965 has worked a farm there. An advocate of conservation and sustainable agriculture, he champions the virtues of the agricultural life for the individual, the community, our country, and our planet.*

*Berry taught English for many years in California and New York and at the University of Kentucky. A fellow of the Guggenheim and Rockefeller Foundations, he won both the T. S. Eliot Award and an award from the National Institute and Academy of Arts and Letters.*

*His more than thirty books of essays and poems include* Home Economics; The Unsettling of America: Culture and Agriculture; The Gift of Good Land; Another Turn of the Crank; Recollected Essays: 1965–1980; *and* Sex, Economy, Freedom, and Community. *He has also edited (with Wes Jackson and Bruce Colman)* Meeting the Expectations of the Land.

*Much of Berry's poetry was written during the 1960s and 1970s, beginning with his publication of* The Broken Ground *in 1964 and ending with* The Wheel *in 1982. Jeffrey A. Lamoureux, a postdoctoral fellow at Duke University, describes Berry's poetry as "resonating with the simple, pastoral values of rural America. His poems [describe] . . . the gritty, honest relationship farmers have with their livestock and the earth. It's good old fashioned blue-collar lyric at its best."*

*A writer in the* New York Review of Books *called Berry "perhaps the great moral essayist of our day." The selection that follows is from* The Unsettling of America, *which Pulitzer Prize–winner Wallace Stegner considered "one of the most cogent books of the decade, and, in its claims for a healthy man-earth relationship, one of our most revolutionary."*

---

From *The Unsettling of America: Culture and Agriculture* by Wendell Berry. Copyright © 1977 by Wendell Berry. Reprinted by permission of Sierra Club Books.

## The Unsettling of America

> *So many goodly citties ransacked and razed; so many nations destroyed and made desolate; so infinite millions of harmelesse people of all sexes, states and ages, massacred, ravaged and put to the sword; and the richest, the fairest and the best part of the world topsiturvied, ruined and defaced for the traffick of Pearles and Pepper: Oh mechanicall victories, oh base conquest.*
>
> —MONTAIGNE

ONE OF THE PECULIARITIES of the white race's presence in America is how little intention has been applied to it. As a people, wherever we have been, we have never really intended to be. The continent is said to have been discovered by an Italian who was on his way to India. The earliest explorers were looking for gold, which was, after an early streak of luck in Mexico, always somewhere farther on. Conquests and foundings were incidental to this search—which did not, and could not, end until the continent was finally laid open in an orgy of goldseeking in the middle of the last century. Once the unknown of geography was mapped, the industrial marketplace became the new frontier, and we continued, with largely the same motives and with increasing haste and anxiety, to displace ourselves—no longer with unity of direction, like a migrant flock, but like the refugees from a broken ant hill. In our own time we have invaded foreign lands and the moon with the high-toned patriotism of the conquistadors, and with the same mixture of fantasy and avarice.

That is too simply put. It is substantially true, however, as a description of the dominant tendency in American history. The temptation, once that has been said, is to ascend altogether into rhetoric and inveigh equally against all our forebears and all present holders of office. To be just, however, it is necessary to remember that there has been another tendency: the tendency to stay put, to say, "No farther. This is the place." So far, this has been the weaker tendency, less glamorous, certainly less successful. It is also the older of these tendencies, having been the dominant one among the Indians.

The Indians did, of course, experience movements of population, but in general their relation to place was based upon old usage and association, upon inherited memory, tradition, veneration. The land was their homeland. The first and greatest American revolution, which has never been superseded, was the coming of people who did *not* look upon the land as a homeland. But there were always those among the newcomers who saw that they had come to a good place and who saw its domestic possibilities. Very early, for instance, there were men who wished to establish agricultural settlements rather than quest for gold or exploit the Indian trade. Later, we know that every advance of the frontier left behind families and communities who intended to remain and prosper where they were.

But we know also that these intentions have been almost systematically overthrown. Generation after generation, those who intended to remain and prosper where they were have been dispossessed and driven out, or subverted and exploited where they were, by those who were carrying out some version of the search for El Dorado. Time after time, in place after place, these conquerors have fragmented and demolished traditional communities, the beginnings of domestic cultures. They have always said that what they destroyed was outdated, provincial, and contemptible. And with alarming frequency they have been believed and trusted by their victims, especially when their victims were other white people.

If there is any law that has been consistently operative in American history, it is that the members of any *established* people or group or community sooner or later become "redskins"—that is, they become the designated victims of an utterly ruthless, officially sanctioned and subsidized exploitation. The colonists who drove off the Indians came to be intolerably exploited by their imperial governments. And that alien imperialism was thrown off only to be succeeded by a domestic version of the same thing; the class of independent small farmers who fought the war of independence has been exploited by, and recruited into, the industrial society until by now it is almost extinct. Today, the most numerous heirs of the farmers of Lexington and Concord are the little groups scattered all over the country whose names begin with "Save": Save Our Land, Save the Valley, Save Our Mountains, Save Our Streams, Save Our Farmland. As so often before, these are *designated* victims—people without official sanction, often without official friends, who are struggling to preserve their places, their values, and their lives as they know them and prefer to live them against the agencies of their own government which are using their own tax moneys against them.

The only escape from this destiny of victimization has been to "succeed"—that is, to "make it" into the class of exploiters, and then to remain so specialized and so "mobile" as to be unconscious of the effects of one's life or livelihood. This escape is, of course, illusory, for one man's producer is another's consumer, and even the richest and most mobile will soon find it hard to escape the noxious effluents and fumes of their various public services.

Let me emphasize that I am not talking about an evil that is merely contemporary or "modern," but one that is as old in America as the white man's presence here. It is an intention that was *organized* here almost from the start. "The New World," Bernard DeVoto wrote in *The Course of Empire*, "was a constantly expanding market.... Its value in gold was enormous but it had still greater value in that it expanded and integrated the industrial systems of Europe."

And he continues: "The first belt-knife given by a European to an Indian was a portent as great as the cloud that mushroomed over Hiroshima.... Instantly the man of 6000 B.C. was bound fast to a way of life that had developed seven and a half millennia beyond his own. He began to live better and he began to die."

The principal European trade goods were tools, cloth, weapons, ornaments, novelties, and alcohol. The sudden availability of these things produced a revolution that "affected every aspect of Indian life. The struggle for existence . . . became easier. Immemorial handicrafts grew obsolescent, then obsolete. Methods of hunting were transformed. So were methods—and the purposes—of war. As war became deadlier in purpose and armament a surplus of women developed, so that marriage customs changed and polygamy became common. The increased usefulness of women in the preparation of pelts worked to the same end. . . . Standards of wealth, prestige, and honor changed. The Indians acquired commercial values and developed business cults. They became more mobile. . . .

"In the sum it was cataclysmic. A culture was forced to change much faster than change could be adjusted to. All corruptions of culture produce breakdowns of morale, of communal integrity, and of personality, and this force was as strong as any other in the white man's subjugation of the red man."

I have quoted these sentences from DeVoto because, the obvious differences aside, he is so clearly describing a revolution that did not stop with the subjugation of the Indians, but went on to impose substantially the same catastrophe upon the small farms and the farm communities, upon the shops of small local tradesmen of all sorts, upon the workshops of independent craftsmen, and upon the households of citizens. It is a revolution that is still going on. The economy is still substantially that of the fur trade, still based on the same general kinds of commercial items: technology, weapons, ornaments, novelties, and drugs. The one great difference is that by now the revolution has deprived the mass of consumers of any independent access to the staples of life: clothing, shelter, food, even water. Air remains the only necessity that the average user can still get for himself, and the revolution has imposed a heavy tax on that by way of pollution. Commercial conquest is far more thorough and final than military defeat. The Indian became a redskin, not by loss in battle, but by accepting a dependence on traders that made *necessities* of industrial goods. This is not merely history. It is a parable.

DeVoto makes it clear that the imperial powers, having made themselves willing to impose this exploitive industrial economy upon the Indians, could not then keep it from contaminating their own best intentions: "More than four-fifths of the wealth of New France was furs, the rest was fish, and it had no agricultural wealth. One trouble was that whereas the crown's imperial policy required it to develop the country's agriculture, the crown's economy required the colony's furs, an adverse interest." And La Salle's dream of developing Louisiana (agriculturally and otherwise) was frustrated because "The interest of the court in Louisiana colonization was to secure a bridgehead for an attack on the silver mines of northern Mexico. . . ."

One cannot help but see the similarity between this foreign colonialism and the domestic colonialism that, by policy, converts productive farm, forest, and grazing lands into strip mines. Now, as then, we see the abstract values of an industrial economy preying upon the native productivity of land and people. The fur trade was only the first establishment on this continent of a mentality whose triumph is its catastrophe.

My purposes in beginning with this survey of history are (1) to show how deeply rooted in our past is the mentality of exploitation; (2) to show how fundamentally revolutionary it is; and (3) to show how crucial to our history—hence, to our own minds—is the question of how we will relate to our land. This question, now that the corporate revolution has so determinedly invaded the farmland, returns us to our oldest crisis.

We can understand a great deal of our history—from Cortés' destruction of Tenochtitlán in 1521 to the bulldozer attack on the coalfields four-and-a-half centuries later—by thinking of ourselves as divided into conquerors and victims. In order to understand our own time and predicament and the work that is to be done, we would do well to shift the terms and say that we are divided between exploitation and nurture. The first set of terms is too simple for the purpose because, in any given situation, it proposes to divide people into two mutually exclusive groups; it becomes complicated only when we are dealing with situations in succession—as when a colonist who persecuted the Indians then resisted persecution by the crown. The terms exploitation and nurture, on the other hand, describe a division not only between persons but also within persons. We are all to some extent the products of an exploitive society, and it would be foolish and self-defeating to pretend that we do not bear its stamp.

Let me outline as briefly as I can what seem to me the characteristics of these opposite kinds of mind. I conceive a strip-miner to be a model exploiter, and as a model nurturer I take the old-fashioned idea or ideal of a farmer. The exploiter is a specialist, an expert; the nurturer is not. The standard of the exploiter is efficiency; the standard of the nurturer is care. The exploiter's goal is money, profit; the nurturer's goal is health—his land's health, his own, his family's, his community's, his country's. Whereas the exploiter asks of a piece of land only how much and how quickly it can be made to produce, the nurturer asks a question that is much more complex and difficult: What is its carrying capacity? (That is: How much can be taken from it without diminishing it? What can it produce *dependably* for an indefinite time?) The exploiter wishes to earn as much as possible by as little work as possible; the nurturer expects, certainly, to have a decent living from his work, but his characteristic wish is to work *as well* as possible. The competence of the exploiter is in organization; that of the nurturer is in order—a human order, that is, that accommodates itself both to other order and to mystery. The exploiter typically serves an institution or organization; the nurturer serves land, household, community, place. The exploiter thinks in terms of numbers, quantities, "hard facts"; the nurturer in terms of character, condition, quality, kind.

It seems likely that all the "movements" of recent years have been representing various claims that nurture has to make against exploitation. The women's movement, for example, when its energies are most accurately placed, is arguing the cause of nurture; other times it is arguing the right of women to be exploiters—which men have no *right* to be. The exploiter is clearly the prototype of the "masculine" man—the wheeler-dealer whose "practical" goals require the sacrifice of flesh, feeling, and principle. The nurturer, on the other hand, has always passed with ease across the boundaries of the so-called sexual roles. Of necessity and without apology, the preserver of seed, the planter, becomes midwife and nurse. Breeder is always metamorphosing into brooder and back again. Over and over again, spring after spring, the questing mind, idealist and visionary, must pass through the planting to become nurturer of the real. The farmer, sometimes known as husbandman, is by definition half mother; the only question is how good a mother he or she is. And the land itself is not mother or father only, but both. Depending on crop and season, it is at one time receiver of seed, bearer and nurturer of young; at another, raiser of seed-stalk, bearer and shedder of seed. And in response to these changes, the farmer crosses back and forth from one zone of spousehood to another, first as planter and then as gatherer. Farmer and land are thus involved in a sort of dance in which the partners are always at opposite sexual poles, and the lead keeps changing: the farmer, as seed-bearer, causes growth; the land, as seed-bearer, causes the harvest.

The exploitive always involves the abuse or the perversion of nurture and ultimately its destruction. Thus, we saw how far the exploitive revolution had penetrated the official character when our recent secretary of agriculture remarked that "Food is a weapon." This was given a fearful symmetry indeed when, in discussing the possible use of nuclear weapons, a secretary of defense spoke of "palatable" levels of devastation. Consider the associations that have since ancient times clustered around the idea of food—associations of mutual care, generosity, neighborliness, festivity, communal joy, religious ceremony—and you will see that these two secretaries represent a cultural catastrophe. The concerns of farming and those of war, once thought to be diametrically opposed, have become identical. Here we have an example of men who have been made vicious, not presumably by nature or circumstance, but by their *values*.

Food is *not* a weapon. To use it as such—to foster a mentality willing to use it as such—is to prepare, in the human character and community, the destruction of the sources of food. The first casualties of the exploitive revolution are character and community. When those fundamental integrities are devalued and broken, then perhaps it is inevitable that food will be looked upon as a weapon, just as it is inevitable that the earth will be looked upon as fuel and people as numbers or machines. But character and community—that is, culture in the broadest, richest sense—constitute, just as much as nature, the source of food. Neither nature nor people alone can produce human sustenance, but only the two together, culturally wedded. The poet Edwin Muir said it unforgettably:

> Men are made of what is made,
> The meat, the drink, the life, the corn,
> Laid up by them, in them reborn.
> And self-begotten cycles close
> About our way; indigenous art

> And simple spells make unafraid
> The haunted labyrinths of the heart
> And with our wild succession braid
> The resurrection of the rose.

To think of food as a weapon, or of a weapon as food, may give an illusory security and wealth to a few, but it strikes directly at the life of all.

The concept of food-as-weapon is not surprisingly the doctrine of a Department of Agriculture that is being used as an instrument of foreign political and economic speculation. This militarizing of food is the greatest threat so far raised against the farmland and the farm communities of this country. If present attitudes continue, we may expect government policies that will encourage the destruction, by overuse, of farmland. This, of course, has already begun. To answer the official call for more production—evidently to be used to bait or bribe foreign countries—farmers are plowing their waterways and permanent pastures; lands that ought to remain in grass are being planted in row crops. Contour plowing, crop rotation, and other conservation measures seem to have gone out of favor or fashion in official circles and are practiced less and less on the farm. This exclusive emphasis on production will accelerate the mechanization and chemicalization of farming, increase the price of land, increase overhead and operating costs, and thereby further diminish the farm population. Thus the tendency, if not the intention, of Mr. Butz's confusion of farming and war, is to complete the deliverance of American agriculture into the hands of corporations.

The cost of this corporate totalitarianism in energy, land, and social disruption will be enormous. It will lead to the exhaustion of farmland and farm culture. Husbandry will become an extractive industry; because maintenance will entirely give way to production, the fertility of the soil will become a limited, unrenewable resource like coal or oil.

This may not happen. It *need* not happen. But it is necessary to recognize that it *can* happen. That it can happen is made evident not only by the words of such men as Mr. Butz, but more clearly by the large-scale industrial destruction of farmland already in progress. If it does happen, we are familiar enough with the nature of American salesmanship to know that it will be done in the name of the starving millions, in the name of liberty, justice, democracy, and brotherhood, and to free the world from communism. We must, I think, be prepeared to see, and to stand by, the truth: that the land should not be destroyed for *any* reason, not even for any apparently good reason. We must be prepared to say that enough food, year after year, is possible only for a limited number of people, and that this possibility can be preserved only by the steadfast, knowledgeable *care* of those people. Such "crash programs" as apparently have been contemplated by the Department of Agriculture in recent years will, in the long run, cause more starvation than they can remedy.

Meanwhile, the dust clouds rise again over Texas and Oklahoma. "Snirt" is falling in Kansas. Snow drifts in Iowa and the Dakotas are black with blown soil. The fields lose their humus and porosity, become less retentive of water, depend more on pesticides, herbicides, chemical fertilizers. Bigger tractors become necessary because the compacted soils are harder to work—and their greater weight further compacts the soil. More and bigger machines, more chemical and methodological shortcuts are needed because of the shortage of manpower on the farm—and the problems of overcrowding and unemployment increase in the cities. It is estimated that it now costs (by erosion) two bushels of Iowa topsoil to grow one bushel of corn. It is variously estimated that from five to twelve calories of fossil fuel energy are required to produce one calorie of hybrid corn energy. An official of the National Farmers Union says that "a farmer who earns $10,000 to $12,000 a year typically leaves an estate valued at about $320,000"—which means that when that farm is financed again, either by a purchaser or by an heir (to pay the inheritance taxes), it simply cannot support its new owner and pay for itself. And the *Progressive Farmer* predicts the disappearance of 200,000 to 400,000 farms each year during the next twenty years if the present trend continues.

The first principle of the exploitive mind is to divide and conquer. And surely there has never been a people more ominously and painfully divided than we are—both against each other and within ourselves. Once the revolution of exploitation is under way, statesmanship and craftsmanship are gradually replaced by salesmanship.* Its stock in trade in politics is to sell despotism and avarice as freedom and democracy. In business it sells sham and frustration as luxury and satisfaction. The "constantly expanding market" first opened in the New World by the fur traders is still expanding—no longer so much by expansions of territory or population, but by the calculated outdating, outmoding, and degradation of goods and by the hysterical self-dissatisfaction of consumers that is indigenous to an exploitive economy.

---

*The craft of persuading people to buy what they do not need, and do not want, for more than it is worth.

This gluttonous enterprise of ugliness, waste, and fraud thrives in the disastrous breach it has helped to make between our bodies and our souls. As a people, we have lost sight of the profound communion—even the union—of the inner with the outer life. Confucius said: "If a man have not order within him / He can not spread order about him. . . ." Surrounded as we are by evidence of the disorders of our souls and our world, we feel the strong truth in those words as well as the possibility of healing that is in them. We see the likelihood that our surroundings, from our clothes to our countryside, are the products of our inward life—our spirit, our vision—as much as they are products of nature and work. If this is true, then we cannot live as we do and be as we would like to be. There is nothing more absurd, to give an example that is only apparently trivial, than the millions who wish to live in luxury and idleness and yet be slender and good-looking. We have millions, too, whose livelihoods, amusements, and comforts are all destructive, who nevertheless wish to live in a healthy environment; they want to run their recreational engines in clean, fresh air. There is now, in fact, no "benefit" that is not associated with disaster. That is because power can be disposed morally or harmlessly only by thoroughly unified characters and communities.

What caused these divisions? There are no doubt many causes, complex both in themselves and in their interaction. But pertinent to all of them, I think, is our attitude toward work. The growth of the exploiters' revolution on this continent has been accompanied by the growth of the idea that work is beneath human dignity, particularly any form of hand work. We have made it our overriding ambition to escape work, and as a consequence have debased work until it is only fit to escape from. We have debased the products of work and have been, in turn, debased by them. Out of this contempt for work arose the idea of a nigger: at first some person, and later some thing, to be used to relieve us of the burden of work. If we began by making niggers of people, we have ended by making a nigger of the world. We have taken the irreplaceable energies and materials of the world and turned them into jim-crack "labor-saving devices." We have made of the rivers and oceans and winds niggers to carry away our refuse, which we think we are too good to dispose of decently ourselves. And in doing this to the world that is our common heritage and bond, we have returned to making niggers of people: we have become each other's niggers.

But is work something that we have a right to escape? And can we escape it with impunity? We are probably the first entire people ever to think so. All the ancient wisdom that has come down to us counsels otherwise. It tells us that work is necessary to us, as much a part of our condition as mortality; that good work is our salvation and our joy; that shoddy or dishonest or self-serving work is our curse and our doom. We have tried to escape the sweat and sorrow promised in Genesis—only to find that, in order to do so, we must forswear love and excellence, health and joy.

Thus we can see growing out of our history a condition that is physically dangerous, morally repugnant, ugly. Contrary to the blandishments of the salesmen, it is not particularly comfortable or happy. It is not even affluent in any meaningful sense, because its abundance is dependent on sources that are being rapidly exhausted by its methods. To see these things is to come up against the question: Then what *is* desirable?

One possibility is just to tag along with the fantasists in government and industry who would have us believe that we can pursue our ideals of affluence, comfort, mobility, and leisure indefinitely. This curious faith is predicated on the notion that we will soon develop unlimited new sources of energy: domestic oil fields, shale oil, gasified coal, nuclear power, solar energy, and so on. This is fantastical because the basic cause of the energy crisis is not scarcity; it is moral ignorance and weakness of character. We don't know *how* to use energy, or what to use it *for*. And we cannot restrain ourselves. Our time is characterized as much by the abuse and waste of human energy as it is by the abuse and waste of fossil fuel energy. Nuclear power, if we are to believe its advocates, is presumably going to be well used by the same mentality that has egregiously devalued and misapplied man- and womanpower. If we had an unlimited supply of solar or wind power, we would use that destructively, too, for the same reasons.

Perhaps all of those sources of energy are going to be developed. Perhaps all of them can sooner or later be developed without threatening our survival. But not all of them together can guarantee our survival, and they cannot define what is desirable. We will not find those answers in Washington, D.C., or in the laboratories of oil companies. In order to find them, we will have to look closer to ourselves.

I believe that the answers are to be found in our history: in its until now subordinate tendency of settlement, of domestic permanence. This was the ambition of thousands of immigrants; it is formulated eloquently in some of the letters of Thomas Jefferson; it was the dream of the freed slaves; it was written into law in the Homestead Act of 1862. There are few of us whose families have not at some time been moved to see its vision and to attempt to enact its possiblity. I am talking about the idea that as many as possible should share in the ownership of the land and thus be bound to it by economic interest, by the investment of love and work, by family loyalty, by memory and tradition. How much land this should be is a question, and the answer will vary with geography. The Homestead Act said 160 acres. The freedmen of the 1860s hoped for forty. We know that, particularly in other countries, families have lived decently on far fewer acres than that.

The old idea is still full of promise. It is potent with healing and with health. It has the power to turn each person away from the big-time promising and planning of the government, to confront in himself, in the immediacy of his own circumstances and whereabouts, the question of what methods and ways are best. It proposes an economy of necessities rather than an economy based upon anxiety, fantasy, luxury, and idle wishing. It proposes the independent, free-standing citizenry that Jefferson thought to be the surest safeguard of democratic liberty. And perhaps most important of all, it proposes an agriculture based upon intensive work, local energies, care, and long-living communities—that is, to state the matter from a consumer's point of view: a dependable, long-term food supply.

This is a possibility that is obviously imperiled—by antipathy in high places, by adverse public fashions and attitudes, by the deterioration of our present farm communities and traditions, by the flawed education and the inexperience of our young people. Yet it alone can promise us the continuity of attention and devotion without which the human life of the earth is impossible.

Sixty years ago, in another time of crisis, Thomas Hardy wrote these stanzas:

> Only a man harrowing clods
>     In a slow silent walk
> With an old horse that stumbles and nods
>     Half asleep as they stalk.
>
> Only thin smoke without flame
>     From the heaps of couch-grass;
> Yet this will go onward the same
>     Though Dynasties pass.

Today most of our people are so conditioned that they do not wish to harrow clods either with an old horse or with a new tractor. Yet Hardy's vision has come to be more urgently true than ever. The great difference these sixty years have made is that, though we feel that this work *must* go onward, we are not so certain that it will. But the care of the earth is our most ancient and most worthy and, after all, our most pleasing responsibility. To cherish what remains of it, and to foster its renewal, is our only legitimate hope.

## The Agricultural Crisis as a Crisis of Culture

In my boyhood, Henry County, Kentucky, was not just a rural county, as it still is—it was a *farming* county. The farms were generally small. They were farmed by families who lived not only upon them, but within and *from* them. These families grew gardens. They produced their own meat, milk, and eggs. The farms were highly diversified. The main money crop was tobacco. But the farmers also grew corn, wheat, barley, oats, hay, and sorghum. Cattle, hogs, and sheep were all characteristically raised on the same farms. There were small dairies, the milking more often than not done by hand. Those were the farm products that might have been considered major. But there were also minor products, and one of the most important characteristics of that old economy was the existence of markets for minor products. In those days a farm family could easily market its surplus cream, eggs, old hens, and frying chickens. The power for field work was still furnished mainly by horses and mules. There was still a prevalent pride in workmanship, and thrift was still a forceful social ideal. The pride of most people was still in their homes, and their homes looked like it.

This was by no means a perfect society. Its people had often been violent and wasteful in their use of the land and of each other. Its present ills had already taken root in it. But I have spoken of its agricultural economy of a generation ago to suggest that there were also good qualities indigenous to it that might have been cultivated and built upon.

That they were not cultivated and built upon—that they were repudiated as the stuff of a hopelessly outmoded, unscientific way of life—is a tragic error on the part of the people themselves; and it is a work of monstrous ignorance and irresponsibility on the part of the experts and politicians, who have prescribed, encouraged, and applauded the disintegration of such farming communities all over the country.

In the decades since World War II the farms of Henry County have become increasingly mechanized. Though they are still comparatively diversified, they are less diversified than they used to be. The holdings are larger, the owners are fewer. The land is falling more and more into the hands of speculators and professional people from the cities, who—in spite of all the scientific agricultural miracles—still have much more money than farmers. Because of big technology and big economies, there is more abandoned land in the county than ever before. Many of the better farms are visibly deteriorating, for want of manpower and time and money to maintain them properly. The number of part-time farmers and ex-farmers increases every year. Our harvests depend more and more on the labor of old people and young children. The farm people live less and less from their own produce, more and more from what they buy. The best of them are more worried about money and more overworked than ever before. Among the people as a whole, the focus of interest has largely shifted from the household to the automobile; the ideals of workmanship and thrift have been replaced by the goals of leisure, comfort, and entertainment. For Henry County plays its full part in what Maurice Telleen calls "the world's first broad-based hedonism." The young people expect to leave as soon as they finish high school, and so they are without permanent interest; they are generally not interested in anything that cannot be reached by automobile on a good road. Few of the farmers' children will be able to afford to stay on the farm—perhaps even fewer will wish to do so, for it will cost too much, require too much work and worry, and it is hardly a fashionable ambition.

And nowhere now is there a market for minor produce: a bucket of cream, a hen, a few dozen eggs. One cannot sell milk from a few cows anymore; the law-required equipment is too expensive. Those markets were done away with in the name of sanitation—but, of course, to the enrichment of the large producers. We have always had to have "a good reason" for doing away with small operators, and in modern times the good reason has often been sanitation, for which there is apparently no small or cheap technology. Further historians will no doubt remark upon the inevitable association, with us, between sanitation and filthy lucre. And it is one of the miracles of science and hygiene that the germs that used to be in our food have been replaced by poisons.

In all this, few people whose testimony would have mattered have seen the connection between the "modernization" of agricultural techniques and the disintegration of the culture and the communities of farming—and the consequent disintegration of the structures of urban life. What we have called agricultural progress has, in fact, involved the forcible displacement of millions of people.

I remember, during the fifties, the outrage with which our political leaders spoke of the forced removal of the populations of villages in communist countries. I also remember that at the same time, in Washington, the word on farming was "Get big or get out"—a policy which is still in effect and which has taken an enormous toll. The only difference is that of method: the force used by the communists was military; with us, it has been economic—a "free market" in which the freest were the richest. The attitudes are equally cruel, and I believe that the results will prove equally damaging, not just to the concerns and values of the human spirit, but to the practicalities of survival.

And so those who could not get big have got out—not just in my community, but in farm communities all over the country. But as a social or economic goal, bigness is totalitarian; it establishes an inevitable tendency toward the *one* that will be the biggest of all. Many who got big to stay in are now being driven out by those who got bigger. The aim of bigness implies not one aim that is not socially and culturally destructive.

And this community-killing agriculture, with its monomania of bigness, is not primarily the work of farmers, though it has burgeoned on their weaknesses. It is the work of the institutions of agriculture: the university experts, the bureaucrats, and the "agribusinessmen," who have promoted so-called efficiency at the expense of community (and of real efficiency), and quantity at the expense of quality.

In 1973, 1000 Kentucky dairies went out of business. They were the victims of policies by which we imported dairy products to compete with our own and exported so much grain as to cause a drastic rise in the price of feed. And, typically, an agriculture expert at the University of Kentucky, Dr. John Nicolai, was optimistic about this failure of 1000 dairymen, whose cause he is supposedly being paid—partly with *their* tax money—to serve. They were inefficient producers, he said, and they needed to be eliminated.

He did not say—indeed, there was no indication that he had ever considered—what might be the limits of his criterion or his logic. Did he propose to applaud this process year after year until "biggest" and "most efficient" become synonymous with "only"? Did these dairymen have any value not subsumed under the heading of "efficiency"? And who benefited by their failure? Assuming that the benefit reached beyond the more "efficient" (that is, the bigger) producers to lower the cost of milk to consumers, do we then have a formula by which to determine how many consumer dollars are equal to the livelihood of one dairyman? Or is *any* degree of "efficiency" worth *any* cost? I do not think that this expert knows the answers. I do not think that he is under any pressure—scholarly, professional, moral, or otherwise—to ask the questions. This sort of regardlessness is invariably justified by pointing to the enormous productivity of American agriculture. But any abundance, in any amount, is illusory if it does not safeguard its producers, and in American agriculture it is now virtually the accepted rule that abundance will destroy its producers.

And along with the rest of society, the established agriculture has shifted its emphasis, and its interest, from quality to quantity, having failed to see that in the long run the two ideas are inseparable. To pursue quantity alone is to destroy those disciplines in the producer that are the only assurance of quantity. What is the effect on quantity of persuading a producer to produce an inferior product? What, in other words, is the relation of pride or craftsmanship to abundance? That is another question the "agribusinessmen" and their academic collaborators do not ask. They do not ask it because they are afraid of the answer: The preserver of abundance is excellence.

My point is that food is a cultural product; it cannot be produced by technology alone. Those agriculturalists who think of the problems of food production solely in terms of technological innovation are oversimplifying both the practicalities of production and the network of meanings and values necessary to define, nurture, and preserve the practical motivations. That the discipline of agriculture should have been so divorced from other disciplines has its immediate cause in the compartmental structure of the universities, in which complementary, mutually sustaining and enriching disciplines are divided, according to "professions," into fragmented, one-eyed specialties. It is suggested, both by the organization of the universities and by the kind of thinking they foster, that farming shall be the responsibility only of the college of agriculture, that law shall be in the sole charge of the professors of law, that morality shall be taken care of by the philosophy department, reading by the English department, and so on. The same, of course, is true of government, which has become another way of institutionalizing the same fragmentation.

However, if we conceive of a culture as one body, which it is, we see that all of its disciplines are everybody's business, and that the proper university product is therefore not the whittled-down, isolated mentality of expertise, but a mind competent in all its concerns. To such a mind it would be clear that there are agricultural disciplines that have nothing to do with crop production, just as there are agricultural obligations that belong to people who are not farmers.

A culture is not a collection of relics or ornaments, but a practical necessity, and its corruption invokes calamity. A healthy culture is a communal order of memory, insight, value, work, conviviality, reverence, aspiration. It reveals the human necessities and the human limits. It clarifies our inescapable bonds to the earth and to each other. It assures that the necessary restraints are observed, that the necessary work is done, and that it is done well. A healthy *farm* culture can be based only upon familiarity and can grow only among a people soundly established upon the land; it nourishes and safeguards a human intelligence of the earth that no amount of technology can satisfactorily replace. The growth of such a culture was once a strong possibility in the farm communities of this country. We now have only the sad remnants of those communities. If we allow another generation to pass without doing what is necessary to enhance and embolden the possibility now perishing with them, we will lose it altogether. And then we will not only invoke calamity—we will deserve it.

Several years ago I argued with a friend of mine that we might make money by marketing some inferior lambs. My friend thought for a minute and then he said, "I'm in the business of producing *good* lambs, and I'm not going to sell any other kind." He also said that he kept the weeds out of his crops for the same reason that he washed his face. The human race has survived by that attitude. It can survive *only* by that attitude—though the farmers who have it have not been much acknowledged or much rewarded.

Such an attitude does not come from technique or technology. It does not come from education; in more than two decades in universities I have rarely seen it. It does not come even from principle. It comes from a passion that is culturally prepared—a passion for excellence and order that is handed down to young people by older people whom they respect and love. When we destroy the possibility of that succession, we will have gone far toward destroying ourselves.

It is by the measure of culture, rather than economics or technology, that we can begin to reckon the nature and the cost of the country-to-city migration that has left our farmland in the hands of only five percent of the people. From a cultural point of view, the movement from the farm to the city involves a radical simplification of mind and of character.

A competent farmer is his own boss. He has learned the disciplines necessary to go ahead on his own, as required by economic obligation, loyalty to his place, pride in his work. His workdays require the use of long experience and practiced judgment, for the failures of which he knows that he will suffer. His days do not begin and end by rule, but in response to necessity, interest, and obligation. They are not measured by the clock, but by the task and his endurance; they last as long as necessary or as long as he can work. He has mastered intricate formal patterns in ordering his work within the overlapping cycles—human and natural, controllable and uncontrollable—of the life of a farm.

Such a man, upon moving to the city and taking a job in industry, becomes a specialized subordinate, dependent upon the authority and judgment of other people. His disciplines are no longer implicit in his own experience, assumptions, and values, but are imposed on him from the outside. For a complex responsibility he has substituted a simple dutifulness. The strict competences of independence, the formal mastery, the complexities of attitude and know-how necessary to life on the farm, which have been in the making in the race of farmers since before history, all are replaced by the knowledge of some fragmentary task that may be learned by rote in a little while.

Such a simplification of mind is easy. Given the pressure of economics and social fashion that has been behind it and the decline of values that has accompanied it, it may be said to have been gravity-powered. The reverse movement—a reverse movement *is* necessary, and some have undertaken it—is uphill, and it is difficult. It cannot be fully accomplished in a generation. It will probably require several generations—enough to establish complex local cultures with strong communal memories and traditions of care.

There seems to be a rule that we can simplify our minds and our culture only at the cost of an oppressive social and mechanical complexity. We can simplify our society—that is, make ourselves free—only by undertaking tasks of great mental and cultural complexity. Farming, the *best* farming, is a task that calls for this sort of complexity, both in the character of the farmer and in his culture. To simplify either one is to destroy it.

That is because the best farming requires a farmer—a husbandman, a nurturer—not a technician or businessman. A technician or a businessman, given the necessary abilities and ambitions, can be made in a little while, by training. A good farmer, on the other hand, is a cultural product; he is made by a sort of training, certainly, in what his time imposes or demands, but he is also made by generations of experience. This essential experience can only be accumulated, tested, preserved, handed down in settled households, friendships, and communities that are deliberately and carefully native to their own ground, in which the past has prepared the present and the present safeguards the future.

The concentration of the farmland into larger and larger holdings and fewer and fewer hands—with the consequent increase of overhead, debt, and dependence on machines—is thus a matter of complex significance, and its agricultural significance cannot be disentangled from its cultural significance. It *forces* a profound revolution in the farmer's mind: once his investment in land and machines is large enough, he must forsake the values of husbandry and assume those of finance and technology. Thenceforth his thinking is not determined by agricultural responsibility, but by financial accountability and the capacities of his machines. Where his money comes from becomes less important to him than where it is going. He is caught up in the drift of energy and interest away from the land. Production begins to override maintenance. The economy of money has infiltrated and subverted the economies of nature, energy, and the human spirit. The man himself has become a consumptive machine.

For some time now ecologists have been documenting the principle that "you can't do one thing"—which means that in a natural system whatever affects one thing ultimately affects everything. Everything in the Creation is related to everything else and dependent on everything else. The Creation is one; it is a uni-verse, a whole, the parts of which are all "turned into one."

A good agricultural system, which is to say a durable one, is similarly unified. In the 1940s, the great British agricultural scientist, Sir Albert Howard, published *An Agricultural Testament* and *The Soil and Health*, in which he argued against the influence in agriculture of "the laboratory hermit" who had substituted "that dreary principle [official organization] for the soul-shaking principle of that essential freedom needed by the seeker after truth." And Howard goes on to speak of the disruptiveness of official organization: "The natural universe, which is one, has been halved, quartered, fractioned.... Real organization always involves real responsibility: the official organization of research tries to retain power and avoid responsibility by sheltering behind groups of experts." Howard himself began as a laboratory hermit: "I could not take my own advice before offering it to other people." But he saw the significance of the "wide chasm between science in the laboratory and practice in the field." He devoted his life to bridging that chasm. His is the story of a fragmentary intelligence seeking both its own wholeness and that of the world. The aim that he finally realized in his books was to prepare the way "for treating the whole problem of health in soil, plant, animal, and man as one great subject." He unspecialized his vision, in other words, so as to see the necessary unity of the concerns of agriculture, as well as the convergence of these concerns with concerns of other kinds: biological, historical, medical, moral, and so on. He sought to establish agriculture upon the same unifying cycle that preserves health, fertility, and renewal in nature: the Wheel of Life (as he called it, borrowing the term from religion), by which "Death supersedes life and life rises again from what is dead and decayed."

It remains only to say what has often been said before—that the best human cultures also have this unity. Their concerns and enterprises are not fragmented, scattered out, at variance or in contention with one another. The people and their work and their country are members of each other and of the culture. If a culture is to hope for any considerable longevity, then the relationships within it must, in recognition of their interdependence, be predominantly cooperative rather than competitive. A people cannot live long at each other's expense or at the expense of their cultural birthright—just as an agriculture cannot live long at the expense of its soil or its work force, and just as in a natural system the competitions among species must be limited if all are to survive.

In any of these systems, cultural or agricultural or natural, when a species or group exceeds the principle of usufruct (literally, the "use of the fruit"), it puts itself in danger. Then, to use an economic metaphor, it is living off the principal rather than the interest. It has broken out of the system of nurture and has become exploitive; it is destroying what gave it life and what it depends upon to live. In all of these systems a fundamental principle must be the protection of the source: the seed, the food species, the soil, the breeding stock, the old and the wise, the keepers of memories, the records.

And just as competition must be strictly curbed within these systems, it must be strictly curbed *among* them. An agriculture cannot survive long at the expense of the natural systems that support it and that provide it with models. A culture cannot survive long at the expense either of its agricultural or of its natural sources. To live at the expense of the source of life is obviously suicidal. Though we have no choice but to live at the expense of other life, it is necessary to recognize the limits and dangers involved: past a certain point in a unified system, "other life" is our own.

The definitive relationships in the universe are thus not competitive but interdependent. And from a human point of view they are analogical. We can build one system only within another. We can have agriculture only within nature, and culture only within agriculture. At certain critical points these systems have to conform with one another or destroy one another.

Under the discipline of unity, knowledge and morality come together. No longer can we have that paltry "objective" knowledge so prized by the academic specialists. To know anything at all becomes a moral predicament. Aware that there is no such thing as a specialized—or even an entirely limitable or controllable—effect, one becomes responsible for judgments as well as facts. Aware that as an agricultural scientist he had "one great subject," Sir Albert Howard could no longer ask, What can I do with what I know? without at the same time asking, How can I be responsible for what I know?

And it is within unity that we see the hideousness and destructiveness of the fragmentary—the kind of mind, for example, that can introduce a production machine to increase "efficiency" without troubling about its effect on workers, on the product, and on consumers; that can accept and even applaud the "obsolescence" of the small farm and not hesitate over the possible political and cultural effects; that can recommend continuous tillage of huge monocultures, with massive use of chemicals and no animal manure or humus, and worry not at all about the deterioration or loss of soil. For cultural patterns of responsible cooperation we have substituted this moral ignorance, which is the etiquette of agricultural "progress."

# Animal Liberation and Vegetarianism 3

PETER SINGER

*Peter Singer is the most well-known living philosopher in the English-speaking world. Known for his clear writing on controversial topics such as neonatal euthanasia, the personhood of late-stage Alzheimer's victims, famine relief, and the environment, Singer has never shied away from truth because of controversy.*

*In 1973, his book* Animal Liberation *created a new movement in the world dedicated to freeing nonhuman animals from second-class status in factory farms, medical experiments, and zoos. The following selection is from this book, where Singer argues that the most practical thing a person can do to liberate animals is to not eat their meat.*

*Another of Singer's books,* Practical Ethics, *had a galvanizing effect on the field of applied ethics by showing how a philosopher could write clear, practical essays urging direct action on everyday topics of real concern. Both this book and* Animal Liberation *were translated into dozens of languages and each sold hundreds of thousands of copies.*

*Singer began his career at Oxford University and was appointed to a professorship at the age of thirty in the Centre for Human Bioethics at Monash University in Australia, where he wrote and taught for two decades. He served as president of the International Association of Bioethics and as editor of its official journal,* Bioethics. *In 1998, Singer was appointed DeCamp Professor in the Center for Human Values at Princeton University. This appointment sparked controversy and numerous articles in English and American magazines and newspapers.*

NOW THAT WE HAVE understood the nature of speciesism and seen the consequences it has for nonhuman animals it is time to ask: What can we do about it? There are many things that we can and should do about speciesism. We should, for instance, write to our political represen-

---

From Peter Singer, *Animal Liberation* (New York: Avon Books, 1990). Reprinted by permission of the author.

tatives about the issues discussed in this book; we should make our friends aware of these issues; we should educate our children to be concerned about the welfare of all sentient beings; and we should protest publicly on behalf of nonhuman animals whenever we have an effective opportunity to do so.

While we should do all these things, there is one other thing we can do that is of supreme importance; it underpins, makes consistent, and gives meaning to all our other activities on behalf of animals. This one thing is that we take responsibility for our own lives, and make them as free of cruelty as we can. The first step is that we cease to eat animals. Many people who are opposed to cruelty to animals draw the line at becoming a vegetarian. It was of such people that Oliver Goldsmith, the eighteenth-century humanitarian essayist, wrote: "They pity, and they eat the objects of their compassion."[1]

As a matter of strict logic, perhaps, there is no contradiction in taking an interest in animals on both compassionate and gastronomic grounds. If one is opposed to inflicting suffering on animals, but not to the painless killing of animals, one could consistently eat animals who had lived free of all suffering and been instantly, painlessly slaughtered. Yet practically and psychologically it is impossible to be consistent in one's concern for nonhuman animals while continuing to dine on them. If we are prepared to take the life of another being merely in order to satisfy our taste for a particular type of food, then that being is no more than a means to our end. In time we will come to regard pigs, cattle, and chickens as things for us to use, no matter how strong our compassion may be; and when we find that to continue to obtain supplies of the bodies of these animals at a price we are able to pay it is necessary to change their living conditions a little, we will be unlikely to regard these changes too critically. The factory farm is nothing more than the application of technology to the idea that animals are means to our ends. Our eating habits are dear to us and not easily altered. We have a strong interest in convincing ourselves that our concern for other animals does not require us to stop eating them. No one in the habit of eating an animal can be completely without bias in judging whether the conditions in which that animal is reared cause suffering.

It is not practically possible to rear animals for food on a large scale without inflicting considerable suffering. Even if intensive methods are not used, traditional farming involves castration, separation of mother and young, breaking up social groups, branding, transportation to the slaughterhouse, and finally slaughter itself. It is difficult to imagine how animals could be reared for food without these forms of suffering. Possibly it could be done on a small scale, but we could never feed today's huge urban populations with meat raised in this manner. If it could be done at all, the animal flesh thus produced would be vastly more expensive than animal flesh is today—and rearing animals is already an expensive and inefficient way of producing protein. The flesh of animals reared and killed with equal consideration for the welfare of animals while they were alive would be a delicacy available only to the rich.

All this is, in any case, quite irrelevant to the immediate question of the ethics of our daily diet. Whatever the theoretical possibilities of rearing animals without suffering may be, the fact is that the meat available from butchers and supermarkets comes from animals who were not treated with any real consideration at all while being reared. So we must ask ourselves, not: Is it *ever* right to eat meat? but: Is it right to eat *this* meat? Here I think that those who are opposed to the needless killing of animals and those who oppose only the infliction of suffering must join together and give the same, negative answer.

Becoming a vegetarian is not merely a symbolic gesture. Nor is it an attempt to isolate oneself from the ugly realities of the world, to keep oneself pure and so without responsibility for the cruelty and carnage all around. Becoming a vegetarian is a highly practical and effective step one can take toward ending both the killing of nonhuman animals and the infliction of suffering upon them. Assume, for the moment, that it is only the suffering that we disapprove of, not the killing. How can we stop the use of the intensive methods of animal rearing described in the previous chapter?

So long as people are prepared to buy the products of intensive farming, the usual forms of protest and political action will never bring about a major reform. Even in supposedly animal-loving Britain, although the wide controversy stirred by the publication of Ruth Harrison's *Animal Machines* forced the government to appoint a group of impartial experts (the Brambell committee) to investigate the issue of mistreatment of animals and make recommendations, when the committee reported the government refused to carry out its recommendations. In 1981 the House of Commons Agriculture Committee made yet another inquiry into intensive farming, and this inquiry also led to recommendations for eliminating the worst abuses. Once again, nothing was done.[2] If this was the fate of the movement for reform in Britain, nothing better can be expected in the United States, where the agribusiness lobby is still more powerful.

This is not to say that the normal channels of protest and political action are useless and should be abandoned. On the contrary, they are a necessary part of the overall struggle for effective change in the treatment of animals. In Britain, especially, organizations like Compassion in World Farming have kept the issue before the public, and even succeeded in bringing about an end to veal crates. More recently American groups have also started to arouse public concern over intensive farming. But in themselves, these methods are not enough.

The people who profit by exploiting large numbers of animals do not need our approval. They need our money. The purchase of the corpses of the animals they rear is the main support the factory farmers ask from the public (the other, in many countries, is big government subsidies). They will use intensive methods as long as they can sell what they produce by these methods; they will have the resources needed to fight reform politically; and they will be able to defend themselves against criticism with the reply that they are only providing the public with what it wants.

Hence the need for each one of us to stop buying the products of modern animal farming—even if we are not convinced that it would be wrong to eat animals who have lived pleasantly and died painlessly. Vegetarianism is a form of boycott. For most vegetarians the boycott is a permanent one, since once they have broken away from flesh-eating habits they can no longer approve of slaughtering animals in order to satisfy the trivial desires of their palates. But the moral obligation to boycott the meat available in butcher shops and supermarkets today is just as inescapable for those who disapprove only of inflicting suffering, and not of killing. Until we boycott meat, and all other products of animal factories, we are, each one of us, contributing to the continued existence, prosperity, and growth of factory farming and all the other cruel practices used in rearing animals for food.

It is at this point that the consequences of speciesism intrude directly into our lives, and we are forced to attest personally to the sincerity of our concern for nonhuman animals. Here we have an opportunity to do something, instead of merely talking and wishing the politicians would do something. It is easy to take a stand about a remote issue, but speciesists, like racists, reveal their true nature when the issue comes nearer home. To protest about bullfighting in Spain, the eating of dogs in South Korea, or the slaughter of baby seals in Canada while continuing to eat eggs from hens who have spent their lives crammed into cages, or veal from calves who have been deprived of their mothers, their proper diet, and the freedom to lie down with their legs extended, is like denouncing apartheid in South Africa while asking your neighbors not to sell their houses to blacks.

To make the boycott aspect of vegetarianism more effective, we must not be shy about our refusal to eat flesh. Vegetarians in omnivorous societies are always being asked about the reasons for their strange diets. This can be irritating, or even embarrassing, but it also provides opportunities to tell people about cruelties of which they may be unaware. (I first learned of the existence of factory farming from a vegetarian who took the time to explain to me why he wasn't eating the same food I was.) If a boycott is the only way to stop cruelty, then we must encourage as many as possible to join the boycott. We can only be effective in this if we ourselves set the example.

People sometimes attempt to justify eating flesh by saying that the animal was already dead when they bought it. The weakness of this rationalization—which I have heard used, quite seriously, many times—should be obvious as soon as we consider vegetarianism as a form of boycott. The nonunion grapes available in stores during the grape boycott inspired by Cesar Chavez's efforts to improve the wages and conditions of the grape-pickers had already been produced by underpaid laborers, and we could no more raise the pay those laborers had received for picking those grapes than we could bring our steak back to life. In both cases the aim of the boycott is not to alter the past but to prevent the continuation of the conditions to which we object.

I have emphasized the boycott element of vegetarianism so much that the reader may ask whether, if the boycott does not spread and prove effective, anything has been achieved by becoming a vegetarian. But we must often venture when we cannot be certain of success, and it would be no argument against becoming a vegetarian if this were all that could be said against it, since none of the great movements against oppression and injustice would have existed if their leaders had made no efforts until they were assured of success. In the case of vegetarianism, however, I believe we do achieve something by our individual acts, even if the boycott as a whole should not succeed. George Bernard Shaw once said that he would be followed to his grave by numerous sheep, cattle, pigs, chickens, and a whole shoal of fish, all grateful at having been spared from slaughter because of his vegetarian diet. Although we cannot identify any individual animals whom we have benefited by becoming a vegetarian, we can assume that our diet, together with that of the many others who are already avoiding meat, will have some impact on the number of animals raised in factory farms and slaughtered for food. This assumption is reasonable because the number of animals raised and slaughtered depends on the profitability of this process, and this profit depends in part on the demand for the product. The smaller the demand, the lower the price and the lower the profit. The lower the profit, the fewer the animals that will be raised and slaughtered. This is elementary economics, and it can easily be observed in tables published by the poultry trade journals, for instance, that there is a direct correlation between the price of poultry and the number of chickens placed in broiler sheds to begin their joyless existence.

So vegetarianism is really on even stronger ground than most other boycotts or protests. The person who boycotts South African produce in order to bring down apartheid achieves nothing unless the boycott succeeds in forcing white South Africans to modify their policies (though the effort may have been well worth making, whatever the outcome); but vegetarians know that they do, by their actions, contribute to a reduction in the suffering and slaughter of animals, whether or not they live to see their efforts spark off a mass boycott of meat and an end to cruelty in farming.

In addition to all this, becoming a vegetarian has a special significance because the vegetarian is a practical, living refutation of a common, yet utterly false, defense of factory farming methods. It is sometimes said that these methods are needed to feed the world's soaring population. Because the truth here is so important—important enough, in fact, to amount to a convincing case for vegetarianism that is quite independent of the question of animal welfare that I have emphasized in this book—I shall digress briefly to discuss the fundamentals of food production.

At this moment, millions of people in many parts of the world do not get enough to eat. Millions more get a sufficient quantity, but they do not get the right kind of food; mostly, they do not get enough protein. The question is, does raising food by the methods practiced in the affluent nations make a contribution to the solution of the hunger problem?

Every animal has to eat in order to grow to the size and weight at which it is considered ready for human beings to eat. If a calf, say, grazes on rough pasture land that grows only grass and could not be planted with corn or any other crop that provides food edible by human beings, the result will be a net gain of protein for human beings, since the grown calf provides us with protein that we cannot—yet—extract economically from grass. But if we take that same calf and place him in a feedlot, or any other confinement system, the picture changes. The calf must now be fed. No matter how little space he and his companions are crowded into, land must be used to grow the corn, sorghum, soybeans, or whatever it is that the calf eats. Now we are feeding the calf food that we ourselves could eat. The calf needs most of this food for the ordinary physiological processes of day-to-day living. No matter how severely the calf is prevented from exercising, his body must still burn food merely to keep him alive. The food is also used to build inedible parts of the calf's body, like bones. Only the food left over after these needs are satisfied can be turned into flesh, and eventually be eaten by human beings.

How much of the protein in his food does the calf use up, and how much is available for human beings? The answer is surprising. It takes twenty-one pounds of protein fed to a calf to produce a single pound of animal protein for humans. We get back less than 5 percent of what we put in. No wonder that Frances Moore Lappé has called this kind of farming "a protein factory in reverse"![3]

We can put the matter another way. Assume we have one acre of fertile land. We can use this acre to grow a high-protein plant food, like peas or beans. If we do this, we will get between three hundred and five hundred pounds of protein from our acre. Alternatively we can use our acre to grow a crop that we feed to animals, and then kill and eat the animals. Then we will end up with between forty and fifty-five pounds of protein from our acre. Interestingly enough, although most animals convert plant protein into animal protein more efficiently than cattle do—a pig, for instance, needs "only" eight pounds of protein to produce one pound for humans—this advantage is almost eliminated when we consider how much protein we can produce per acre, because cattle can make use of sources of protein that are indigestible for pigs. So most estimates conclude that plant foods yield about ten times as much protein per acre as meat does, although estimates vary, and the ratio sometimes goes as high as twenty to one.[4]

If instead of killing the animals and eating their flesh we use them to provide us with milk or eggs we improve our return considerably. Nevertheless the animals must still use protein for their own purposes and the most efficient forms of egg and milk production do not yield more than a quarter of the protein per acre that can be provided by plant foods.

Protein is, of course, only one necessary nutrient. If we compare the total number of calories produced by plant foods with animal foods, the comparison is still all in favor of plants. A comparison of yields from an acre sown with oats or broccoli with yields from an acre used for feed to produce pork, milk, poultry, or beef shows that the acre of oats produces six times the calories yielded by pork, the most efficient of the animal products. The acre of broccoli yields nearly three times as many calories as pork. Oats produce more than twenty-five times as many calories per acre as beef. Looking at some other nutrients shatters other myths fostered by meat and dairy industries. For instance, an acre of broccoli produces twenty-four times the iron produced by an acre used for beef, and an acre of oats sixteen times the same amount of iron. Although milk production does yield more calcium per acre than oats, broccoli does better still, providing five times as much calcium as milk.[5]

The implications of all this for the world food situation are staggering. In 1974 Lester Brown of the Overseas Development Council estimated that if Americans were to reduce their meat consumption by only 10 percent for one year, it would free at least 12 million tons of grain for human consumption—or enough to feed 60 million people. Don Paarlberg, a former U.S. assistant secretary of agriculture, has said that merely reducing the U.S. livestock population by half would make available enough food to make up the calorie deficit of the nonsocialist underdeveloped nations nearly four times over.[6] Indeed, the food wasted by animal production in the affluent nations would be sufficient, if properly distributed, to end both hunger and malnutrition throughout the world. The simple answer to our question, then, is that raising animals for food by the methods used in the industrial nations does not contribute to the solution of the hunger problem.

Meat production also puts a strain on other resources. Alan Durning, a researcher at the Worldwatch Institute, an environmental thinktank based in Washington, D.C., has calculated that one pound of steak from steers raised in a feedlot costs five pounds of grain, 2,500 gallons of water, the energy equivalent of a gallon of gasoline, and about thirty-five pounds of eroded topsoil. More than a third of North America is taken up with grazing, more than half of U.S. croplands are planted with livestock feed, and more than half of all water consumed in the United States goes to livestock.[7] In all these respects plant foods are far less demanding of our resources and our environment.

Let us consider energy usage first. One might think that agriculture is a way of using the fertility of the soil and the energy provided by sunlight to increase the amount of energy available to us. Traditional agriculture does precisely that. Corn grown in Mexico, for instance, produces 83 calories of food for each calorie of fossil fuel energy input. Agriculture in developed countries, however, relies on a large input of fossil fuel. The most energy-efficient form of food production in the United States (oats, again) produces barely 2.5 food calories per calorie of fossil fuel energy, while potatoes yield just over 2, and wheat and soybeans around 1.5. Even these meager results, however, are a bonanza compared to United States animal production, every form of which costs more energy than it yields. The least inefficient—rangeland beef—uses more than 3 calories of fossil fuel for every food calorie it yields; while the most inefficient—feedlot beef—takes 33 fuel calories for every food calorie. In energy efficiency, eggs, lamb, milk, and poultry come between the two forms of beef production. In other words, limiting ourselves to United States agriculture, growing crops is generally at least five times more energy-efficient than grazing cattle, about twenty times more energy-efficient than producing chickens, and more than fifty times as energy-efficient as feedlot cattle production.[8] United States animal production is workable only because it draws on millions of years of accumulated solar energy, stored in the ground as oil and coal. This makes economic sense to agribusiness corporations because meat is worth more than oil; but for a rational long-term use of our finite resources, it makes no sense at all.

Animal production also compares poorly with crop production as far as water use is concerned. A pound of meat requires fifty times as much water as an equivalent quantity of wheat.[9] *Newsweek* graphically described this volume of water when it said, "The water that goes into a 1000 pound steer would float a destroyer."[10] The demands of animal production are drying up the vast underground pools of water on which so many of the drier regions of America, Australia, and other countries rely. In the cattle country that stretches from western Texas to Nebraska, for example, water tables are falling and wells are going dry as the huge underground lake known as the Ogalalla Aquifer—another resource which, like oil and coal, took millions of years to create—continues to be used up to produce meat.[11]

Nor should we neglect what animal production does to the water that it does not use. Statistics from the British Water Authorities Association show that there were more than 3,500 incidents of water pollution from farms in 1985. Here is just one example from that year: a tank at a pig unit burst, sending a quarter-million liters of pig excrement into the River Perry and killing 110,000 fish. More than half of the prosecutions by water authorities for serious pollution of rivers are now against farmers.[12] This is not surprising, for a modest 60,000-bird egg factory produces eighty-two tons of manure every week, and in the same period two thousand pigs will excrete twenty-seven tons of manure and thirty-two tons of urine. Dutch farms produce 94 million tons of manure a year, but only 50 million can safely be absorbed by the land. The excess, it has been calculated, would fill a freight train stretching 16,000 kilometers from Amsterdam to the farthest shores of Canada. But the excess is not being carted away; it is dumped on the land where it pollutes water supplies and kills the remaining natural vegetation in the farming regions of the Netherlands.[13] In the United States, farm animals produce 2 billion tons of manure a year—about ten times that of the human population—and half of it comes from factory-reared animals, where the waste does not return naturally to the land.[14] As one pig farmer put it: "Until fertilizer gets more expensive than labor, the waste has very little value to me."[15] So the manure that should restore the fertility of our soils ends up polluting our streams and rivers.

It will, however, be the squandering of the forests that turns out to be the greatest of all the follies caused by the demand for meat. Historically, the desire to graze animals has been the dominant motive for clearing forests. It still is today. In Costa Rica, Colombia, and Brazil, in Malaysia, Thailand, and Indonesia, rainforests are being cleared to provide grazing land for cattle. But the meat produced from the cattle does not benefit the poor of those countries. Instead it is sold to the well-to-do in the big cities, or it is exported. Over the past twenty-five years, nearly half of Central America's tropical rainforests have been destroyed, largely to provide beef to North America.[16] Perhaps 90 percent of the plant and animal species on this planet live in the tropics, many of them still unrecorded by science.[17] If the clearing continues at its present rate, they will be pushed into extinction. In addition, clearing the land causes erosion, the increased runoff leads to flooding, peasants no longer have wood for fuel, and rainfall may be reduced.[18]

We are losing these forests just at the moment when we are starting to learn how truly vital they are. Since the North American drought of 1988, many people have heard of the threat posed to our planet by the greenhouse effect, caused mainly by increasing amounts of carbon dioxide in the atmosphere. Forests store immense amounts of carbon; it has been estimated that despite all the clearing that has taken place, the world's remaining forests still hold four hundred times the amount of carbon released into the atmosphere each year by human use of fossil fuels. Destroying a forest releases the carbon into the atmosphere in the form of carbon dioxide. Conversely, a new, growing forest absorbs carbon dioxide from the atmosphere and locks it up as living matter. The destruction of existing forests will intensify the greenhouse effect; in large-scale reforestation, combined with other measures to reduce the output of carbon dioxide, lies our only hope of mitigating it.[19] If we fail to do so, the warming of our planet will mean, within the next fifty years, widespread droughts, further destruction of forests from climatic change, the extinction of innumerable species unable to cope with the changes in their habitat, and a melting of the polar ice caps, which will in turn raise sea levels and flood coastal cities and plains. A rise of one meter in the level of the sea would flood 15 percent of Bangladesh, affecting 10 million people; and it would threaten the very existence of some low-lying Pacific island nations such as the Maldives, Tuvalu, and Kiribati.[20]

Forests and meat animals compete for the same land. The prodigious appetite of the affluent nations for meat means that agribusiness can pay more than those who want to preserve or restore the forests. We are, quite literally, gambling with the future of our planet—for the sake of hamburgers.

How far should we go? The case for a radical break in our eating habits is clear; but should we eat nothing but plant foods? Where exactly do we draw the line?

Drawing precise lines is always difficult. I shall make some suggestions, but the reader might well find what I say here less convincing than what I have said before about the more clear-cut cases. You must decide for yourself where you are going to draw the line, and your decision may not coincide exactly with mine. This does not matter all that much. We can distinguish bald men from men who are not bald without deciding every borderline case. It is agreement on the fundamentals that is important.

I hope that anyone who has read this far will recognize the moral necessity of refusing to buy or eat the flesh or other products of animals who have been reared in modern factory farm conditions. This is the clearest case of all, the absolute minimum that anyone with the capacity to look beyond considerations of narrow self-interest should be able to accept.

Let us see what this minimum involves. It means that, unless we can be sure of the origin of the particular item we are buying, we must avoid chicken, turkey, rabbit, pork, veal, beef, and eggs. At the present time relatively little lamb is intensively produced; but some is, and more may be in future. The likelihood of your beef coming from a feedlot or some other form of confinement—or from grazing land created by clearing rainforest—will depend on the country in which you live. It is possible to obtain supplies of all these meats that do not come from factory farms, but unless you live in a rural area this takes a lot of effort. Most butchers have no idea how the animals whose bodies they are selling were raised. In some cases, such as that of chickens, traditional methods of rearing have disappeared so completely that it is almost impossible to buy a chicken that was free to roam outdoors; and veal is a meat that simply cannot be produced humanely. Even when meat is described as "organic" this may mean no more than that the animals were not fed the usual doses of antibiotics, hormones, or other drugs; small solace for an animal who was not free to walk around outdoors. As for eggs, in many countries "free range eggs" are widely available, though in most parts of the United States they are still very difficult to get.

Once you have stopped eating poultry, pork, veal, beef, and factory farm eggs the next step is to refuse to eat any slaughtered bird or mammal. This is only a very small additional step, since so few of the birds or mammals commonly eaten are not intensively reared. People who have no experience of how satisfying an imaginative vegetarian diet can be may think of it as a major sacrifice. To this I can only say: "Try it!" Buy a good vegetarian cookbook and you will find that being a vegetarian is no sacrifice at all. The reason for taking this extra step may be the belief that it is wrong to kill these creatures for the trivial purpose of pleasing our palates; or it may be the knowledge that even when these animals are not intensively raised they suffer in the various other ways described in the previous chapter.

Now more difficult questions arise. How far down the evolutionary scale shall we go? Shall we eat fish? What about shrimps? Oysters? To answer these questions we must bear in mind the central principle on which our concern for other beings is based. As I said in the first chapter, the only legitimate boundary to our concern for the interests of other beings is the point at which it is no longer accurate to say that the other being has interests. To have interests, in a strict, nonmetaphorical sense, a being must be capable of suffering or experiencing pleasure. If a being suffers, there can be no moral justification for disregarding that suffering, or for refusing to count it equally with the like suffering of any other being. But the converse of this is also true. If a being is not capable of suffering, or of enjoyment, there is nothing to take into account.

So the problem of drawing the line is the problem of deciding when we are justified in assuming that a being is incapable of suffering. In my earlier discussion of the evidence that nonhuman animals are capable of suffering, I suggested two indicators of this capacity: the behavior of the being, whether it writhes, utters cries, attempts to escape from the source of pain, and so on; and the similarity of the nervous system of the being to our own. As we proceed down the evolutionary scale we find that on both these grounds the strength of the evidence for a capacity to feel pain diminishes. With birds and mammals the evidence is overwhelming. Reptiles and fish have nervous systems that differ from those of mammals in some important respects but share the basic structure of centrally organized nerve pathways. Fish and reptiles show most of the pain behavior that mammals do. In most species there is even vocalization, although it is not audible to our ears. Fish, for instance, make vibratory sounds, and different "calls" have been distinguished by researchers, including sounds indicating "alarm" and "aggravation."[21] Fish also show signs of distress when they are taken out of the water and allowed to flap around in a net or on dry land until they die. Surely it is only because fish do not yelp or whimper in a way that we can hear that otherwise decent people can think it a pleasant way of spending an afternoon to sit by the water dangling a hook while previously caught fish die slowly beside them.

In 1976 the British Royal Society for the Prevention of Cruelty to Animals set up an independent panel of inquiry into shooting and angling. The panel was chared by Lord Medway, a noted zoologist, and made up of experts outside the RSPCA. The inquiry examined in detail evidence on whether fish can feel pain, and concluded unequivocally that the evidence for pain in fish is as strong as the evidence for pain in other vertebrate animals.[22] People more concerned about causing pain than about killing may ask: Assuming fish *can* suffer, how much *do* they actually suffer in the normal process of commercial fishing? It may seem that fish, unlike birds and mammals, are not made to suffer in the process of rearing them for our tables, since they are usually not reared at all: human beings interfere with them only to catch and kill them. Actually this is not always true: fish farming—which is as intensive a form of factory farming as raising feedlot beef—is a rapidly growing industry. It began with freshwater fish like trout, but the Norwegians developed a technique for producing salmon in cages in the sea, and other countries are now using this method for a variety of marine fish. The potential welfare problems of farmed fish, such as stocking densities, the denial of the migratory urge, stress during handling, and so on have not even been investigated. But even with fish who are not farmed, the death of a commercially caught fish is much more drawn out than the death of, say, a chicken, since fish are simply hauled up into the air and left to die. Since their gills can extract oxygen from water but not from air, fish out of the water cannot breathe. The fish on sale in your supermarket may have died slowly, from suffocation. If it was a deep-sea fish, dragged to the surface by the net of a trawler, it may have died painfully from decompression.

When fish are caught rather than farmed, the ecological argument against eating intensively reared animals does not apply to fish. We do not waste grain or soybeans by feeding them to fish in the ocean. Yet there is a different ecological argument that counts against the extensive commercial fishing of the oceans now practiced, and this is that we are rapidly fishing out the oceans. In recent years fish catches have declined dramatically. Several once-abundant species of fish, such as the herrings of Northern Europe, the California sardines, and the New England haddock, are now so scarce as to be, for commercial purposes, extinct. Modern fishing fleets trawl the fishing grounds systematically with fine-gauge nets that catch everything in their way. The nontarget species—known in the industry as "trash"—may make up as much as half the catch.[23] Their bodies are thrown overboard. Because trawling involves dragging a huge net along the previously undisturbed bottom of the ocean, it damages the fragile ecology of the seabed. Like other ways of producing animal food, such fishng is also wasteful of fossil fuels, consuming more energy than it produces.[24] The nets used by the tuna fishing industry, moreover, also catch thousands of dolphins every year, trapping them underwater and drowning them. In addition to the disruption of ocean ecology caused by all this overfishing there are bad consequences for humans too. Throughout the world, small coastal villages that live by fishing are finding their traditional source of food and income drying up. From the communities of Ireland's west coast to the Burmese and Malayan fishing villages the story is the same. The fishing industry of the developed nations has become one more form of redistribution from the poor to the rich.

So out of concern for both fish and human beings we should avoid eating fish. Certainly those who continue to eat fish while refusing to eat other animals have taken a major step away from speciesism; but those who eat neither have gone one step further.

When we go beyond fish to the other forms of marine life commonly eaten by humans, we can no longer be quite so confident about the existence of a capacity for pain. Crustacea—lobster, crabs, prawns, shrimps—have nervous systems very different from our own. Nevertheless, Dr. John Baker, a zoologist at the University of Oxford and a fellow of the Royal Society, has stated that their sensory organs are highly developed, their nervous systems complex, their nerve cells very similar to our own, and their responses to certain stimuli immediate and vigorous. Dr. Baker therefore believes that lobster, for example, can feel pain. He is also clear that the standard method of killing lobster—dropping them into boiling water—can cause pain for as long as two minutes. He experimented with other methods sometimes said to be more humane, such as putting them in cold water and heating them slowly, or leaving them in fresh water until they cease to move, but found that both of these led to more prolonged struggling and, apparently, suffering.[25] If crustacea can suffer, there must be a great deal of suffering involved, not only in the method by which they are killed, but also in the ways in which they are transported and kept alive at markets. To keep them fresh they are frequently simply packed, alive, on top of each other. So even if there is some room for doubt about the capacity of these animals to feel pain, the fact that they may be suffering a great deal, combined with the absence of any need to eat them on our part, makes the verdict plain: they should receive the benefit of the doubt.

Oysters, clams, mussels, scallops, and the like are mollusks, and mollusks are in general very simple organisms. (There is an exception: the octopus is a mollusk, but far more developed, and presumably more sentient, than its distant mollusk relatives.) With creatures like oysters, doubts about a capacity for pain are considerable; and in the first edition of this book I suggested that somewhere between a shrimp and an oyster seems as good a place to draw the line as any. Accordingly, I continued occasionally to eat oysters, scallops, and mussels for some time after I became, in every other respect, a vegetarian. But while one cannot with any confidence say that these creatures do feel pain, so one can equally have little confidence in saying that they do not feel pain. Moreover, if they do feel pain, a meal of oysters or mussels would inflict pain on a considerable number of creatures. Since it is so easy to avoid eating them, I now think it better to do so.[26]

This takes us to the end of the evolutionary scale, so far as creatures we normally eat are concerned; essentially, we are left with a vegetarian diet. The traditional vegetarian diet, however, includes animal products, such as eggs and milk. Some have tried to accuse vegetarians of inconsistency here. "Vegetarian," they say, is a word that has the same root as "vegetable" and a vegetarian should eat only food of vegetable origin. Taken as a verbal quibble, this criticism is historically inaccurate. The term "vegetarian" came into general use as a result of the formation of the Vegetarian Society in England in 1847. Since the rules of the society permit the use of eggs and milk, the term "vegetarian" is properly applied to those who use these animal products. Recognizing this linguistic *fait accompli*, those who eat neither animal flesh nor eggs nor milk nor foods made from milk call themselves "vegans." The verbal point, however, is not the important one. What we should ask is whether the use of these other animal products is morally justifiable. This question is a real one because it is possible to be adequately nourished without consuming any animal products at all—a fact that is not widely known, although most people now know that vegetarians can live long and healthy lives. I shall say more on the topic of nutrition later in this chapter; for the present it is enough to know that we can do without eggs and milk. But is there any reason why we should?

We have seen that the egg industry is one of the most ruthlessly intensive forms of modern factory farming, exploiting hens relentlessly to produce the most eggs at the least cost. Our obligation to boycott this type of farming is as strong as our obligation to boycott intensively produced pork or chicken. But what of free-range eggs, assuming you can get them? Here the ethical objections are very much less. Hens provided with both shelter and an outdoor run to walk and scratch around in live comfortably. They do not appear to mind the removal of their eggs. The main grounds for objection are that the male chicks of the egg-laying strain will have been killed on hatching, and the hens themselves will be killed when they cease to lay productively. The question is, therefore, whether the pleasant lives of the hens (plus the benefits to us of the eggs) are sufficient to outweigh the killing that is a part of the system. One's answer to that will depend on one's view about killing, as distinct from the infliction of suffering.[27] In keeping with the reasons given there, I do not, on balance, object to free-range egg production.

Milk and milk products like cheese and yogurt raise different issues. We have seen in Chapter 3 [of Singer's *Animal Liberation*] that dairy production can be distressing for the cows and their calves in several ways: the necessity of making the cow pregnant, and the subsequent separation of the cow and her calf; the increasing degree of confinement on many farms; the health and stress problems caused by feeding cows very rich diets and breeding them for ever-greater milk yields; and now the prospect of further stress from daily injections of bovine growth hormone.

In principle, there is no problem in doing without dairy products. Indeed, in many parts of Asia and Africa, the only milk ever consumed is human milk, for infants. Many adults from these parts of the world lack the ability to digest the lactose that milk contains, and they become ill if they drink milk. The Chinese and Japanese have long used soybeans to make many of the things we make from dairy products. Soy milks are now widely available in Western countries, and tofu ice cream is popular with those trying to reduce their intake of fat and cholesterol. There are even cheeses, spreads, and yogurts made from soybeans.

Vegans, then, are right to say that we ought not to use dairy products. They are living demonstrations of the practicality and nutritional soundness of a diet that is totally free from the exploitation of other animals. At the same time, it should be said that, in our present speciesist world, it is not easy to keep so strictly to what is morally right. A reasonable and defensible plan of action is to change your diet at a measured pace with which you can feel comfortable. Although in principle all dairy products are replaceable, in practice in Western societies it is much more difficult to cut out meat and dairy products than it is to eliminate meat alone. Until you start reading food labels with an eye to avoiding dairy products, you will never believe how many foods contain them. Even buying a tomato sandwich becomes a problem, since it will probably be spread either with butter, or with a margarine containing whey or nonfat milk. There is little gained for animals if you give up animal flesh and battery eggs, and simply replace them with an increased amount of cheese. On the other hand, the following is, if not ideal, a reasonable and practical strategy:

- replace animal flesh with plant foods;
- replace factory farm eggs with free-range eggs if you can get them; otherwise avoid eggs;
- replace the milk and cheese you buy with soymilk, tofu, or other plant foods, but do not feel obliged to go to great lengths to avoid all food containing milk products.

Eliminating speciesism from one's dietary habits is very difficult to do all at once. People who adopt the strategy I support here have made a clear public commitment to the movement against animal exploitation. The most urgent task of the Animal Liberation movement is to persuade as many people as possible to make this commitment, so that the boycott will spread and gain attention. If because of an admirable desire to stop all forms of exploitation of animals immediately we convey the impression that unless one gives up milk products one is no better than those who still eat animal flesh, the result may be that many people are deterred from doing anything at all, and the exploitation of animals will continue as before.

These, at least, are some of the answers to problems that are likely to face nonspeciesists who ask what they should and should not eat. As I said at the beginning of this section, my remarks are intended to be no more than suggestions. Sincere nonspeciesists may well disagree among themselves about the details. So long as there is agreement on the fundamentals this should not disrupt efforts toward a common goal.

\* \* \*

Many people are willing to admit that the case for vegetarianism is strong. Too often, though, there is a gap between intellectual conviction and the action needed to break a lifetime habit. There is no way in which books can bridge this gap; ultimately it is up to each one of us to put our convictions into practice. But I can try, in the next few pages, to narrow the gap. My aim is to make the transition from an omnivorous diet to a vegetarian one much easier and more attractive, so that instead of seeing the change of diet as an unpleasant duty the reader looks forward to a new and interesting cuisine, full of fresh foods as well as unusual meatless dishes from Europe, China, and the Middle East, dishes so varied as to make the habitual meat, meat, and more meat of most Western diets stale and repetitive by comparison. The enjoyment of such a cuisine is enhanced by the knowledge that its good taste and nourishing qualities were provided directly by the earth, neither wasting what the earth produces, nor requiring the suffering and death of any sentient being.

Vegetarianism brings with it a new relationship to food, plants, and nature. Flesh taints our meals. Disguise it as we may, the fact remains that the centerpiece of our dinner has come to us from the slaughterhouse, dripping blood. Untreated and unrefrigerated, it soon begins to putrefy and stink. When we eat it, it sits heavily in our stomachs, blocking our digestive processes until, days later, we struggle to excrete it.[28] When we eat plants, food takes on a different quality. We take from the earth food that is ready for us and does not fight against us as we take it. Without meat to deaden the palate we experience an extra delight in fresh vegetables taken straight from the ground. Personally, I found the idea of picking my own dinner so satisfying that shortly after becoming a vegetarian I began digging up part of our backyard and growing some of my own vegetables—something that I had never thought of doing previously, but that several of my vegetarian friends were also doing. In this way dropping flesh-meat from my diet brought me into closer contact with plants, the soil, and the seasons.

Cooking, too, was something I became interested in only after I became a vegetarian. For those brought up on the usual Anglo-Saxon menus, in which the main dish consists of meat supplemented by two overcooked vegetables, the elimination of meat poses an interesting challenge to the imagination. When I speak in public about the issues discussed in this book, I am often asked about what one can eat instead of meat, and it is clear from the way the question is phrased that the questioner has mentally subtracted the chop or hamburger from his or her plate, leaving the mashed potatoes and boiled cabbage, and is wondering what to put in place of the meat. A heap of soybeans perhaps?

There may be those who would enjoy such a meal, but for most tastes the answer is to rethink the entire idea of the main course, so that it consists of a combination of ingredients, perhaps with a salad on the side, instead of detached items. Good Chinese dishes, for instance, are superb combinations of one or more high-protein ingredients—in vegetarian Chinese cooking, they may include tofu, nuts, bean sprouts, mushrooms, or wheat gluten, with fresh, lightly cooked vegetables and rice. An Indian curry using lentils for protein, served over brown rice with some fresh sliced cucumber for light relief, makes an equally satisfying meal, as does an Italian vegetarian lasagna with salad. You can even make "tofu meatballs" to put on top of your spaghetti. A simpler meal might consist of whole grains and vegetables. Most Westerners eat very little millet, whole wheat, or buckwheat, but these grains can form the basis of a dish that is a refreshing change. . . . Some people find it hard, at first, to change their attitude to a meal. Getting used to meals without a central piece of animal flesh may take time, but once it has happened you will have so many interesting new dishes to choose from that you will wonder why you ever thought it would be difficult to do without flesh foods.

Apart from the tastiness of their meals, people contemplating vegetarianism are most likely to worry about whether they will be adequately nourished. These worries are entirely groundless. Many parts of the world have vegetarian cultures whose members have been as healthy, and often healthier, than nonvegetarians living in similar areas. Strict Hindus have been vegetarians for more than two thousand years. Gandhi, a lifelong vegetarian, was close to eighty when an assassin's bullet ended his active life. In Britain, where there has now been an official vegetarian movement for more than 140 years, there are third- and fourth-generation vegetarians. Many prominent vegetarians, such as Leonardo da Vinci, Leo Tolstoy, and George Bernard Shaw, have lived long, immensely creative lives. Indeed, most people who have reached exceptional old age have eaten little or no meat. The inhabitants of the Vilcabamba valley in Ecuador frequently live to be more than one hundred years old, and men as old as 123 and 142 years have been found by scientists; these people eat less than one ounce of meat a week. A study of all living centenarians in Hungary found that they were largely vegetarian.[29] That meat is unnecessary for physical endurance is shown by a long list of successful athletes who do not eat it, a list that includes Olympic long-distance swimming champion Murray Rose, the famous Finnish distance runner Paavo Nurmi, basketball star Bill Walton, the "ironman" triathlete Dave Scott, and 400-meter Olympic hurdle champion Edwin Moses.

Many vegetarians claim that they feel fitter, healthier, and more zestful than when they ate meat. A great deal of new evidence now supports them. The 1988 United States Surgeon General's Report on Nutrition and Health cites a major study indicating that the death rate for heart attacks of vegetarians between the ages of thirty-five and sixty-four is only 28 percent of the rate for Americans in general in that age group. For older vegetarians the rate of death from heart attacks was still less than half that of nonvegetarians. The same study showed that vegetarians who ate eggs and dairy products had cholesterol levels 16 percent lower than those of meat eaters, and vegans had cholesterol levels 29 percent lower. The report's main recommendations were to reduce consumption of cholesterol and fat (especially saturated fat), and increase consumption of whole grain foods and cereal products, vegetables (including dried beans and peas) and fruits. A recommendation to reduce cholesterol and saturated fat is, in effect, a recommendation to avoid meat (except perhaps chicken from which the skin has been removed), and cream, butter, and all except low-fat dairy products.[30] The report was widely criticized for failing to be more specific in saying this—a vagueness due, apparently, to successful lobbying by groups like the National Cattlemen's Association and the Dairy Board.[31] Whatever lobbying took place failed, however, to prevent the section on cancer from reporting that studies have found an association between breast cancer and meat intake, and also between eating meat, especially beef, and cancer of the large bowel. The American Heart Association has also been recommending, for many years, that Americans reduce their meat intake.[32] Diets designed for health and longevity like the Pritikin plan and the McDougall plan are either largely or entirely vegetarian.[33]

Nutritional experts no longer dispute about whether animal flesh is essential; they now agree that it is not. If ordinary people still have misgivings about doing without it, these misgivings are based on ignorance. Most often this ignorance is about the nature of protein. We are frequently told that protein is an important element in a sound diet, and that meat is high in protein. Both these statements are true, but there are two other things that we are told less often. The first is that the average American eats too much protein. The protein intake of the average American exceeds the generous level recommended by the National Academy of Sciences by 45 percent. Other estimates say that most Americans consume between two and four times as much meat as the body can use. Excess protein cannot be stored. Some of it is excreted, and some may be converted by the body to carbohydrate, which is an expensive way to increase one's carbohydrate intake.[34]

The second thing to know about protein is that meat is only one among a great variety of foods containing protein, its chief distinction being that it is the most expensive. It was once thought that meat protein was of superior quality, but as long ago as 1950 the British Medical Association's Committee on Nutrition stated:

> It is generally accepted that it is immaterial whether the essential protein units are derived from plant or animal foods, provided that they supply an appropriate mixture of the units in assimilable form.[35]

More recent research has provided further confirmation of this conclusion. We now know that the nutritional value of protein consists in the essential amino acids it contains, since these determine how much of the protein the body can use. While it is true that animal foods, especially eggs and milk, have a very well-balanced amino acid composition, plant foods like soybeans and nuts also contain a broad range of these nutrients. Moreover by eating different kinds of plant proteins at the same time it is easy to put together a meal that provides protein entirely equivalent to that of animal protein. This principle is called "protein complementarity," but you do not need to know much about nutrition to apply it. The peasant who eats his beans or lentils with rice or corn is practicing protein complementarity. So is the mother who gives her child a peanut butter sandwich on whole wheat bread—a combination of peanuts and wheat, both of which contain protein. The different forms of protein in the different foods combine with each other in such a way that the body absorbs more protein if they are eaten together than if they were eaten separately. Even without the complementary effect of combining different proteins, however, most of the plant foods we eat—not just nuts, peas, and beans, but even wheat, rice, and potatoes—contain enough protein in themselves to provide our bodies with the protein we need. If we avoid junk foods that are high in sugar or fats and nothing else, about the only way we can fail to get enough protein is if we are on a diet that is insufficient in calories.[36]

Protein is not the only nutrient in meat, but the others can all easily be obtained from a vegetarian diet without special care. Only vegans, who take no animal products at all, need to be especially careful about their diet. There appears to be one, and only one, necessary nutrient that is not normally available from plant sources, and this is vitamin B12, which is present in eggs and milk, but not in a readily assimilable form in plant foods. It can, however, be obtained from seaweeds such as kelp, from a soy sauce made by the traditional Japanese fermentation method, or from tempeh, a fermented soybean product eaten in parts of Asia, and often now available in health food stores in the West. It is also possible that it is produced by microorganisms in our own intestines. Studies of vegans who have not taken any apparent source of B12 for many years have shown their blood levels of this vitamin still to be within the normal range. Nevertheless to make sure of avoiding a deficiency, it is simple and inexpensive to take vitamin tablets containing B12. The B12 in these tablets is obtained from bacteria grown on plant foods. Studies of children in vegan families have shown that they develop normally on diets that contain a B12 supplement but no animal food after weaning.[37]

I have tried in this chapter to answer the doubts about becoming a vegetarian that can easily be articulated and expressed. But some people have a deeper resistance that makes them hesitate. Perhaps the reason for hesitation is a fear of being thought a crank by one's friends. When my wife and I began to think about becoming vegetarians we talked about this. We worried that we would be cutting ourselves off from our nonvegetarian friends and at that time none of our long-established friends was vegetarian. The fact that we became vegetarians together certainly made the decision easier for both of us, but as things turned out we need not have worried. We explained our decision to our friends and they saw that we had good reasons for it. They did not all become vegetarians, but they did not cease to be our friends either; in fact I think they rather enjoyed inviting us to dinner and showing us how well they could cook without meat. Of course, it is possible that you will encounter people who consider you a crank. This is much less likely now than it was a few years ago, because there are so many more vegetarians. But if it should happen, remember that you are in good company. All the best reformers—those who first opposed the slave trade, nationalistic wars, and the exploitation of children working a fourteen-hour day in the factories of the Industrial Revolution—were at first derided as cranks by those who had an interest in the abuses they were opposing.

## Notes

1. Oliver Goldsmith, *The Citizen of the World*, in *Collected Works*, ed. A. Friedman (Oxford: Clarendon Press, 1966), vol. 2, p. 60. Apparently Goldsmith himself fell into this category, however, since according to Howard Williams in *The Ethics of Diet* (abridged edition, Manchester and London, 1907, p. 149), Goldsmith's sensibility was stronger than his self-control.

2. In attempting to rebut the argument for vegetarianisn presented in this chapter of the first edition, R. G. Frey described the reforms proposed by the House of Commons Agriculture Committee in 1981, and wrote: "The House of Commons as a whole has yet to decide on this report, and it may well be in the end that it is diluted; but even so, there is no doubt that it represents a significant advance in combating the abuses of factory farming." Frey then argued that the report showed that these abuses could be overcome by tactics that stopped short of advocating a boycott of animal products. (R. G. Frey, *Rights, Killing and Suffering*, Oxford: Blackwell, 1983, pp. 207.) This is one instance in which I sincerely wish my critic had been right; but the House of Commons did not bother to "dilute" the report of its Agriculture Committee—it simply ignored it. Eight years later nothing has changed for the vast majority of Britain's intensively produced animals. The exception is for veal calves, where a consumer boycott *did* play a significant role.

3. Frances Moore Lappé, *Diet for a Small Planet* (New York: Friends of the Earth/Ballantine, 1971), pp. 4–11. This book is the best brief introduction to the topic, and figures not otherwise attributed in this section have been taken from this book. (A revised edition was published in 1982.) The main original sources are *The World Food Problem*, a Report of the President's Science Advisory Committee (1967); *Feed Situation*, February 1970, U.S. Department of Agriculture; and *National and State Livestock-Feed Relationships*, U.S. Department of Agriculture, Economic Research Service, Statistical Bulletin No. 446, February 1970.

4. The higher ratio is from Folke Dovring, "Soybeans," *Scientific American*, February 1974. Keith Akers presents a different set of figures in *A Vegetarian Sourcebook* (New York: Putnam, 1983), chapter 10. His tables compare per-acre nutritional returns from oats, broccoli, pork, milk, poultry, and beef. Even though oats and broccoli are not high-protein foods, none of the animal foods produced even half as much protein as the plant foods. Akers's original sources are: United States Department of Agriculture, *Agricultural Statistics*, 1979; United States Department of Agriculture, *Nutritive Value of American Foods* (Washington, D.C., U.S. Government Printing Office, 1975); and C. W. Cook, "Use of Rangelands for Future Meat Production," *Journal of Animal Science* 45: 1476 (1977).

5. Keith Akers, *A Vegetarian Sourcebook*, pp. 90–91, using the sources cited above.

6. Boyce Rensberger, "Curb on U.S. Waste Urged to Help World's Hungry," *The New York Times*, October 25, 1974.

7. *Science News*, March 5, 1988, p. 153, citing *Worldwatch*, January/February 1988.

8. Keith Akers, *A Vegetarian Sourcebook*, p. 100, based on D. Pimental and M. Pimental, *Food, Energy and Society* (New York: Wiley, 1979), pp. 56, 59, and U.S. Department of Agricul-

ture, *Nutritive Value of American Foods* (Washington, D.C.: U.S. Government Printing Office, 1975).

9. G. Borgstrom, *Harvesting the Earth* (New York: Abelard-Schuman, 1973), pp. 64–65; cited by Keith Akers, *A Vegetarian Sourcebook*.

10. "The Browning of America," *Newsweek*, February 22, 1981, p. 26; quoted by John Robbins, *Diet for a New America* (Walpole, N.H.: Stillpoint, 1987), p. 367.

11. "The Browning of America," p. 26.

12. Fred Pearce, "A Green Unpleasant Land," *New Scientist*, July 24, 1986, p. 26.

13. Sue Armstrong, "Marooned in a Mountain of Manure," *New Scientist*, November 26, 1988.

14. J. Mason and P. Singer, *Animal Factories* (New York: Crown, 1980), p. 84, citing R. C. Loehr, *Pollution Implications of Animal Wastes—A Forward Oriented Review*, Water Pollution Control Research Series (U.S. Environmental Protection Agency, Washington, D.C., 1968), pp. 26–27; H. A. Jasiorowski, "Intensive Systems of Animal Production," in R. L. Reid, ed., *Proceedings of the II World Conference on Animal Production* (Sydney: Sydney University Press, 1975), p. 384; and J. W. Robbins, *Environmental Impact Resulting from Unconfined Animal Production* (Cincinnati: Environmental Research Information Center, U.S. Environmental Protection Agency, 1978), p. 9.

15. "Handling Waste Disposal Problems," *Hog Farm Management*, April 1978, p. 17, quoted in J. Mason and P. Singer, *Animal Factories*, p. 88.

16. Information from the Rainforest Action Network, *The New York Times*, January 22, 1986, p. 7.

17. E. O. Williams, *Biophilia* (Cambridge: Harvard University Press, 1984), p. 137.

18. Keith Akers, *A Vegetarian Sourcebook*, pp. 99–100; based on H. W. Anderson, et al., *Forests and Water: Effects of Forest Management on Floods, Sedimentation and Water Supply*, U.S. Department of Agriculture Forest Service General Technical Report PSW–18/1976; and J. Kittridge, "The Influence of the Forest on the Weather and Other Environmental Factors," in United Nations Food and Agriculture Organization, *Forest Influences* (Rome, 1962).

19. Fred Pearce, "Planting Trees for a Cooler World," *New Scientist*, October 15, 1988, p. 21.

20. David Dickson, "Poor Countries Need Help to Adapt to Rising Sea Level," *New Scientist*, October 7, 1989, p. 4; Sue Wells and Alasdair Edwards, Gone with the Waves," *New Scientist*, November 11, 1989, pp. 29–32.

21. L. and M. Milne, *The Senses of Men and Animals* (Middlesex and Baltimore: Penguin Books, 1965), chapter 5.

22. *Report of the Panel of Enquiry into Shooting and Angling*, published by the panel in 1980 and available through the Royal Society for the Prevention of Cruelty to Animals (U.K.), paragraphs 15–57.

23. Geoff Maslen, "Bluefin, the Making of the Mariners," *The Age* (Melbourne), January 26, 1985.

24. D. Pimental and M. Pimental, *Food, Energy and Society* (New York: Wiley, 1979), chapter 9; I owe this reference to Keith Akers, *A Vegetarian Sourcebook*, p. 117.

25. See J. R. Baker: *The Humane Killing of Lobsters and Crabs*, The Humane Education Centre, London, no date; J. R. Baker and M. B. Dolan, "Experiments on the Humane Killing of Lobsters and Crabs," *Scientific Papers of the Humane Education Centre* 2: 1–24 (1977).

26. My change of mind about mollusks stems from conversations with R. I. Sikora.

27. See pages 230–231 below.

28. "Struggle" is not altogether a joke. According to a comparative study published in *The Lancet* (December 30, 1972), the "mean transit time" of food through the digestive system of a sample group of nonvegetarians on a Western type of diet was between seventy-six and eighty-three hours; for vegetarians, forty-two hours. The authors suggest a connection between the length of time the stool remains in the colon and the incidence of cancer of the colon and related diseases which have increased rapidly in nations whose consumption of meat has increased but are almost unknown among rural Africans who, like vegetarians, have a diet low in meat and high in roughage.

29. David Davies, "A Shangri-La in Ecuador," *New Scientist*, February 1, 1973. On the basis of other studies, Ralph Nelson of the Mayo Medical School has suggested that a high protein intake causes us to "idle our metabolic engine at a faster rate" (*Medical World News*, November 8, 1974, p. 106). This could explain the correlation between longevity and little or no meat consumption.

30. *The Surgeon General's Report on Nutrition and Health* (Washington, D.C.: U.S. Government Printing Office, 1988).

31. According to a wire-service report cited in *Vegetarian Times*, November 1988.

32. *The New York Times*, October 25, 1974.

33. N. Pritikin and P. McGrady, *The Pritikin Program for Diet and Exercise* (New York: Bantam, 1980); J. J. McDougall, *The McDougall Plan* (Piscataway, N.J.: New Century, 1983).

34. Francis Moore Lappé, *Diet for a Small Planet*, pp. 28–29; see also *The New York Times*, October 25, 1974; *Medical World News*, November 8, 1974, p. 106.

35. Quoted in F. Wokes, "Proteins," *Plant Foods for Human Nutrition*, 1: 38 (1968).

36. In the first edition of *Diet for a Small Planet* (1971), Frances Moore Lappé emphasized protein complementarity to show that a vegetarian diet can provide enough protein. In the revised edition (New York: Ballantine, 1982) this emphasis has disappeared, replaced by a demonstration that a healthy vegetarian diet is bound to contain enough protein even without complementarity. For another account of the adequacy of plant foods as far as protein is concerned, see Keith Akers, *A Vegetarian Sourcebook*, chapter 2.

37. F. R. Ellis and W. M. E. Montegriffo, "The Health of Vegans," *Plant Foods for Human Nutrition*, vol. 2, pp. 93–101 (1971). Some vegans claim that B12 supplements are not necessary, on the grounds that the human intestine can synthesize this vitamin from other B-group vitamins. The question is, however, whether this synthesis takes place sufficiently early in the digestive process for the B12 to be absorbed rather than excreted. At present the nutritional adequacy of an all-plant diet without supplementation is an open scientific question; accordingly it seems safer to take supplementary B12. See also F. Wokes, "Proteins," *Plant Foods for Human Nutrition*, p. 37.

# Meat Is Good for You                                                4

STUART PATTON

*Stuart Patton, Ph.D., worked in food science for more than fifty years, retiring from teaching at Pennsylvania State University in 1981 but continuing bench research until 1999. He published more than 240 articles on food safety and nutrition and supervised the training of thirty-three graduate students. In August 2000, a special symposium was held in Patton's honor at Penn State, which then named an auditorium in his honor and established an endowment in his name. He is now professor emeritus of food science at Pennsylvania State University.*

*For many years, Patton was scientific advisor to the American Council on Science and Health, which (according to its self-description) is a nonprofit, independent organization of scientists, physicians, and food analysts that attempts to give the public a balanced scientific view about food.*

*Patton's editorial from the American Council on Science and Education that follows attacks some of the arguments in the previous article about meat, specifically the alleged health benefits of a vegetarian diet.*

IN RECENT MONTHS, many papers have carried articles and letters critical of meat (particularly beef), milk, and milk products. We are told to reduce or eliminate them from our diet to prevent or cure various diseases and environmental problems. But these prescriptions are scientifically invalid and won't achieve their touted goals. They are urged upon us despite the fact that nutritionists have recognized the exceptional dietary merits of meat and milk for over a century.

With respect to the environment, we are told that animal agriculture wastes land and grain that are needed in programs to feed starving people.

---

Reprinted with permission of the American Council on Science and Health (ACSH). 1995 Broadway, 2nd Floor, New York, NY 10023-5800. Dr. Patton is scientific advisor to ACSH and professor emeritus of food science at Pennsylvania State University. To learn more about ACSH visit www.acsh.org.

But the real challenge is to overcome the political and economic factors that prevent getting enough food to the hungry. The criticism also involves a misperception of animal agriculture, discussed below.

Since the United States formed some 200 years ago, Americans have grown up on meat and milk. At the time of that founding, our European forbears averaged only slightly more than 5 feet in height. In today's terms, this suggests undernourishment. The outstanding value of milk as a source of calcium and the high quality of both meat and milk proteins no doubt have been factors in the notable subsequent increase in Americans' stature, as well as in the improvement in our general health.

The anti–meat-and-milk campaign is consistent with the political agenda of the animal rights movement's strict vegetarianism ideal. They suggest that Americans' risk of cancer will increase as well as that of several other major diseases (heart disease, for example), and that we will be likelier to die young unless we reduce or eliminate these foods from our diets. This is not science; it is propaganda.

Cancer, for example, is a complicated family of diseases. Except for lung cancer, we know very little about the specific causes of most cancers. No one knows just how and to what extent food is involved.

Animal fat has been singled out as a culprit in the development of cancer, yet studies show that very-low-fat diets as well as high-fat diets are associated with a greater incidence of some forms of cancer. In addition, research has demonstrated the presence of cancer antagonists (conjugated linoleic acids, or CLAs) in milk fat. And, despite the dire warnings of the anti–meat-and-milk groups, the latest American Cancer Society statistics indicate that cancer incidence in the U.S., with the exception of lung cancer, is actually falling!

The supposed danger of cholesterol is another of their widely used scares. This accusation stems from the longstanding presumed guilt-by-association of dietary cholesterol with heart disease. The facts: The blood cholesterol values of 80 to 85 percent of people show little or no response to dairy cholesterol.

The criticism by the anti-beef contingent that animal agriculture misdirects the use of land and grain indicates misunderstanding on their part. A major reason that humans have domesticated livestock is that livestock do not compete with mankind for food. Specifically, the natural food of cattle consists of grasses and crude plant matter that humans neither want to eat nor can digest. There is a lot of land for which the best use is animal grazing. In fact, it is not fit for much else.

Grassland farming of these animals is practiced in many areas of the world today, including parts of the U.S. A high-grain regimen for steers improves palatability in certain cuts of meat for which some Americans are willing to pay a premium. But such feeding practices are hardly an indispensable requirement in the meat industry. If anything, the current market, which is demanding leaner meat, tends toward less grain feeding.

The anti–meat-and-milk coalition charges that the dairy and beef industries, backed by the federal government, are guilty of discrimination for promoting the consumption of foods that are harmful to certain races and ethnic groups. For example, the capacity to digest lactose, or milk sugar, varies among races. What are the facts? Lactose is no problem for the majority of Americans and can be easily managed, using several different options, by the rest. Virtually everyone can accommodate a glass of milk a day without intestinal discomfort from the lactose.

Moreover, lactose digestion is controlled by a dominant gene that is spreading in the U.S. population. So lactose intolerance is not only a nonproblem; it is disappearing.

Flat statements that meat and milk products should be reduced or eliminated from the diet are misguided and simplistic. Obviously, the human condition varies greatly and many of us have individual needs based on age, gender, occupation, health status, lifestyle, and genetic background. While satisfactory diets for some individuals may be created without the use of meat or milk, it is a mistake to assume that everyone will know how much of what to put into such a diet to achieve sound nutrition. In this sense, dietary inclusion of meat and milk products is excellent insurance because of their many nutrients.

In any case, it is irresponsible to use one's political agenda in an effort to confuse and mislead the public about the sciences of nutrition and public health.

# Lifeboat Ethics
## The Case against Helping the Poor

GARRETT HARDIN

*Garrett Hardin graduated from the University of Chicago in 1936 and earned a Ph.D. in microbial ecology from Stanford University in 1942. He joined the Biology Department at the University of California–Santa Barbara in 1946. Hardin became an activist in 1963, first for abortion rights, then for the environment. His 1968 essay "The Tragedy of the Commons" catapulted him to fame. Increasing requests for his services led him to reduce his teaching commitment at the university after 1970; he took early retirement in 1978.*

*Hardin has written or edited a dozen books, including* Filters against Folly *(1985). Hardin describes himself as an "eco-conservative," asserting that the society that survives is one that follows those rules that serve its long-term interests.*

*In the following selection, Hardin argues that it is counterproductive to send food to the starving. Appealing to consequentialist and utilitarian premises, Hardin argues that sending food only creates more people in the future who will die, because the problem of limited resources and overpopulation is not addressed by merely sending food.*

LIFEBOAT ETHICS is merely a special application of the logic of the commons.[1] The classic paradigm is that of a pasture held as common property by a community and governed by the following rules: first, each herdsman may pasture as many cattle as he wishes on the commons; and second, the gain from the growth of cattle accrues to the individual owners of the cattle. In an underpopulated world the system of the commons may do no harm and may even be the most economic way to manage things, since management costs are kept to a minimum. In an overpopulated (or overexploited) world a system of the commons leads to ruin, because each

---

Pages 120–37 from George R. Lucas, Jr., and Thomas W. Ogletree, eds., *Lifeboat Ethics: The Moral Dilemmas of World Hunger.* Copyright © 1976 by Vanderbilt University and the Society for Values in Higher Education. Reprinted by permission of HarperCollins Publishers, Inc.

herdsman has more to gain individually by increasing the size of his herd than he has to lose as a single member of the community guilty of lowering the carrying capacity of the environment. Consequently he (with others) overloads the commons.

Even if an individual fully perceives the ultimate consequences of his actions he is most unlikely to act in any other way, for he cannot count on the restraint *his* conscience might dictate being matched by a similar restraint on the part of *all* the others. (Anything less than all is not enough.) Since mutual ruin is inevitable, it is quite proper to speak of the *tragedy* of the commons.

Tragedy is the price of freedom in the commons. Only by changing to some other system (socialism or private enterprise, for example) can ruin be averted. In other words, in a crowded world survival requires that some freedom be given up. (We have, however, a choice in the freedom to be sacrificed.) Survival is possible under several different politico-economic systems—but not under the system of the commons. When we understand this point, we reject the ideal of distributive justice stated by Karl Marx a century ago, "From each according to his ability, to each according to his needs."[2] This ideal might be defensible if "needs" were defined by the larger community rather than by the individual (or individual political unit) *and if "needs" were static.*[3] But in the past quarter-century, with the best will in the world, some humanitarians have been asserting that rich populations must supply the needs of poor populations even though the recipient populations increase without restraint. At the United Nations conference on population in Bucharest in 1973 spokesmen for the poor nations repeatedly said in effect: "We poor people have the right to reproduce as much as we want to; you in the rich world have the responsibility of keeping us alive."

Such a Marxian disjunction of rights and responsibilities inevitably tends toward tragic ruin for all. It is almost incredible that this position is supported by thoughtful persons, but it is. How does this come about? In part, I think, because language deceives us. When a disastrous loss of life threatens, people speak of a "crisis," implying that the threat is temporary. More subtle is the implication of quantitative stability built into the pronoun "they" and its relatives. Let me illustrate this point with quantified prototype statements based on two different points of view.

*Crisis analysis:* "*These* poor people (1,000,000) are starving, because of a crisis (flood, drought, or the like). How can we refuse *them* (1,000,000)? Let us feed *them* (1,000,000). Once the crisis is past those who are still hungry are few (say 1,000) and there is no further need for our intervention."

*Crunch analysis:* "*Those* (1,000,000) who are hungry are reproducing. We send food to *them* (1,010,000). *Their* lives (1,020,000) are saved. But since the environment is still essentially the same, the next year *they* (1,030,000) ask for more food. We send it to *them* (1,045,000); and the next year *they* (1,068,000) ask for still more. Since the need has not gone away, it is a mistake to speak of a passing crisis: it is evidently a permanent crunch that this growing 'they' face—a growing disaster, not a passing state of affairs."

"They" increases in size. Rhetoric makes no allowance for a ballooning pronoun. Thus we can easily be deceived by language. We cannot deal adequately with ethical questions if we ignore quantitative matters. This attitude has been rejected by James Sellers, who dismisses prophets of doom from Malthus[4] to Meadows[5] as "chiliasts." Chiliasts (or millenialists, to use the Latin-derived equivalent of the Greek term) predict a catastrophic end of things a thousand years from some reference point. The classic example is the prediction of Judgment Day in the year 1000 anno Domini. Those who predicted it were wrong, of course; but the fact that this specific prediction was wrong is no valid criticism of the use of numbers in thinking. Millenialism is numerology, not science.

In science, most of the time, it is not so much exact numbers that are important as it is the relative size of numbers and the direction of change in the magnitude of them. Much productive analysis is accomplished with only the crude quantitation of "order of magnitude" thinking. First and second derivatives are often calculated with no finer aim than to find out if they are positive or negative. Survival can hinge on the crude issue of the sign of change, regardless of number. This is a far cry from the spurious precision of numerology. Unfortunately, the chasm between the "two cultures," as C. P. Snow called them,[6] keeps many in the non-scientific culture from understanding the significance of the quantitative approach. One is tempted to wonder also whether an additional impediment to understanding may not be the mortal sin called Pride, which some theologians regard as the mother of all sins.

Returning to Marx, it is obvious that the *each* in "to each according to his needs" is not—despite the grammar—a unitary, stable entity: "each" is a place-holder for a ballooning variable. Before we commit ourselves to saving the life of *each* and every person in need we had better ask this question: "*And then what?*" That is, what about tomorrow, what about posterity? As Hans Jonas has pointed out,[7] traditional ethics has almost entirely ignored the claims of posterity. In an overpopulated world humanity cannot long endure under a regime governed by posterity-blind ethics. It is the essence of ecological ethics that it pays attention to posterity.

Since "helping" starving people requires that we who are rich give up some of our wealth, any refusal to do so is almost sure to be attributed to selfishness. Selfishness there may be, but focusing on selfishness is likely to be non-productive. In truth, a selfish motive can be found in all policy proposals. The selfishness of *not* giving is obvious and need not be elaborated. But the selfishness of giving is no less real, though more subtle.[8] Consider the sources of support for Public Law 480, the act of Congress under which surplus foods were given to poor countries, or sold to them at bargain prices ("concessionary terms" is the euphemism). Why did we give food away? Conventional wisdom says it was because we momentarily transcended our normal selfishness. Is that the whole story?

It is not. The "we" of the above sentence needs to be subdivided. The farmers who grew the grain did not give it away. They sold it to the government (which then gave it away). Farmers received selfish benefits in two ways: the direct sale of grain, and the economic support to farm prices given by this governmental purchase in an otherwise free market. The operation of P. L. 480 during the past quarter-century brought American farmers to a level of prosperity never known before.

Who else benefited—in a selfish way? The stockholders and employees of the railroads that moved grain to seaports benefited. So also did freight-boat operators (U.S. "bottoms" were specified by law). So also did grain elevator operators. So also did agricultural research scientists who were financially supported in a burgeoning but futile effort "to feed a hungry world."[9] And so also did the large bureaucracy required to keep the P. L. 480 system working. In toto, probably several million people personally benefited from the P. L. 480 program. Their labors cannot be called wholly selfless.

Who *did* make a sacrifice for P. L. 480? The citizens generally, nearly two hundred million of them, paying directly or indirectly through taxes. But each of these many millions lost only a little: whereas each of the million or so gainers gained a great deal. The blunt truth is that *philanthropy pays*—if you are hired as a philanthropist. Those on the gaining side of P. L. 480 made a great deal of money and could afford to spend lavishly to persuade Congress to continue the program. Those on the sacrificing side sacrificed only a little bit per capita and could not afford to spend much protecting their pocketbooks against philanthropic inroads. And so P. L. 480 continued, year after year.

Should we condemn philanthropy when we discover that some of its roots are selfish? I think not, otherwise probably no philanthropy would be possible. The secret of practical success in large-scale public philanthropy is this: see to it that the losses are widely distributed so that the per capita loss is small, but concentrate the gains in a relatively few people so that these few will have the economic power needed to pressure the legislature into supporting the program.

I have spent some time on this issue because I would like to dispose once and for all of condemnatory arguments based on "selfishness." As a matter of principle we should always assume that selfishness is *part* of the motivation of every action. But what of it? If Smith proposes a certain public policy, it is far more important to know whether the policy will do public harm or public good than it is to know whether Smith's motives are selfish or selfless. Consequences ("ends") can be more objectively determined than motivations ("means"). Situational ethics wisely uses consequences as the measure of morality. "If the end does not justify the means, what does?" asks Joseph Fletcher.[10] The obsession of older ethical systems with means and motives is no doubt in part a consequence of envy, which has a thousand disguises.[11] (Though I am sure this is true, the situationist should not dwell on envy very long, for it is after all only a motive, and as such not directly verifiable. In any case public policy must be primarily concerned with consequences.)

Even judging an act by its consequences is not easy. We are limited by the basic theorem of ecology, "We can never do merely one thing."[12] The fact that an act has many consequences is all the more reason for de-emphasizing motives as we carry out our ethical analyses. Motives by definition apply only to intended consequences. The multitudinous unintended ones are commonly denigrated by the term "side-effects." But "The road to hell is paved with good intentions," so let's have done with motivational evaluations of public policy.

Even after we have agreed to eschew motivational analysis, foreign aid is a tough nut to crack. The literature is large and contradictory, but it all points to the inescapable conclusion that a quarter of a century of earnest effort has not conquered world poverty. To many observers the theat of future disasters is more convincing now than it was a quarter of a century ago—and the disasters are not all in the future either.[13] Where have we gone wrong in foreign aid?

We wanted to do good, of course. The question, "How can we help a poor country?" seems like a simple question, one that should have a simple answer. Our failure to answer it suggests that the question is not as simple as we thought. The variety of contradictory answers offered is disheartening.

How can we find our way through this thicket? I suggest we take a cue from a mathematician. The great algebraist Karl Jacobi (1804–1851) had a simple stratagem that he recommended to students who found themselves butting their heads against a stone wall. *Umkehren, immer umkehren*—"Invert, always invert." Don't just keep asking the same old question over and over: turn it upside down and ask the opposite question. The answer you get then may not be the one you want, but it may throw useful light on the question you started with.

Let's try a Jacobian inversion of the food/population problem. To sharpen the issue, let us take a particular example, say India. The question we want to answer is, "How can we help India?" But since that approach has repeatedly thrust us against a stone wall, let's pose the Jacobian invert, "How can we *harm* India?" After we've answered this perverse question we will return to the original (and proper) one.

As a matter of method, let us grant ourselves the most malevolent of motives: let us ask, "How can we harm India—*really* harm her?" Of course we might plaster the country with thermonuclear bombs, speedily wiping out most of the 600 million people. But, to the truly malevolent mind, that's not much fun: a dead man is beyond harming. Bacterial warfare could be a bit "better," but not much. No: we want something that will really make India suffer, not merely for a day or a week, but on and on and on. How can we achieve this inhumane goal?

Quite simply: by sending India a bounty of food, year after year. The United States exports about 80 million tons of grain a year. Most of it we sell: the foreign exchange it yields we use for such needed imports as petroleum (38 percent of our oil consumption in 1974), iron ore, bauxite, chromium, tin, etc. But in the pursuit of our malevolent goal let us "unselfishly" tighten our belts, make sacrifices, and do without that foreign exchange. Let us *give* all 80 million tons of grain to the Indians each year.

On a purely vegetable diet it takes about 400 pounds of grain to keep one person alive and healthy for a year. The 600 million Indians need 120 million tons per year; since their nutrition is less than adequate presumably they are getting a bit less than that now. So the 80 million tons we give them will almost double India's per capita supply of food. With a surplus, Indians can afford to vary their diet by growing some less efficient crops; they can also convert some of the grain into meat (pork and chickens for the Hindus, beef and chickens for the Moslems). The entire nation can then be supplied not only with plenty of calories, but also with an adequate supply of high quality protein. The people's eyes will sparkle, their steps will become more elastic; and they will be capable of more work. "Fatalism" will no doubt diminish. (Much so-called fatalism is merely a consequence of malnutrition.) Indians may even become a bit overweight, though they will still be getting only two-thirds as much food as the average inhabitant of a rich country. Surely—we think—surely a well-fed India would be better off?

Not so: *ceteris paribus,* they will ultimately be worse off. Remember, "We can never do merely one thing." A generous gift of food would have not only nutritional consequences: it would also have political and economic consequences. The difficulty of distributing free food to a poor people is well known. Harbor, storage, and transport inadequacies result in great losses of grain to rats and fungi. Political corruption diverts food from those who need it most to those who are more powerful. More abundant supplies depress free market prices and discourage native farmers from growing food in subsequent years. Research into better ways of agriculture is also discouraged. Why look for better ways to grow food when there is food enough already?

There are replies, of sorts, to all the above points. It may be maintained that all these evils are only temporary ones: in time, organizational sense will be brought into the distributional system and the government will crack down on corruption. Realizing the desirability of producing more food, for export if nothing else, a wise government will subsidize agricultural research in spite of an apparent surplus. Experience does not give much support to this optimistic view, but let us grant the conclusions for the sake of getting on to more important matters. Worse is to come.

The Indian unemployment rate is commonly reckoned at 30 percent, but it is acknowledged that this is a minimum figure. *Under*employment is rife. Check into a hotel in Calcutta with four small bags and four bearers will carry your luggage to the room—with another man to carry the key. Custom, and a knowledge of what the traffic will bear, decree this practice. In addition malnutrition justifies it in part. Adequately fed, half as many men would suffice. So one of the early consequences of achieving a higher level of nutrition in the Indian population would be to increase the number of unemployed.

India needs many things that food will not buy. Food will not diminish the unemployment rate (quite the contrary); nor will it increase the supply of minerals, bicycles, clothes, automobiles, gasoline, schools, books, movies, or television. All these things require energy for their manufacture and maintenance.

Of course, food is a form of energy, but it is convertible to other forms only with great loss; so we are practically justified in considering energy and food as mutually exclusive goods. On this basis the most striking difference between poor and rich countries is not in the food they eat but in the energy they use. On a per capita basis rich countries use about three times as much of the primary foods—grain and the like—as do poor countries. (To a large extent this is because the rich convert much of the grain to more "wasteful" animal meat.) But when it comes to energy, rich countries use ten times as much per capita. (Near the extremes Americans use 60 times as much per person as Indians.) By reasonable standards much of this energy may be wasted (e.g., in the manufacture of "exercycles" for sweating the fat off people who have eaten too much), but a large share of this energy supplies the goods we regard as civilized: effortless transportation, some luxury foods, a variety of sports, clean space-heating, more than adequate clothing, and energy-consuming arts—music, visual arts, electronic auxiliaries, etc. Merely giving food to a people does almost nothing to satisfy the appetite for any of these other goods.

But a well-nourished people is better fitted to try to wrest more energy from its environment. The question then is this: Is the native environment able to furnish more energy? And at what cost?

In India energy is already being gotten from the environment at a fearful cost. In the past two centuries millions of acres of India have been deforested in the struggle for fuel, with the usual environmental degradation. The Vale of Kashmir, once one of the garden spots of the world, has been denuded to such an extent that the hills no longer hold water as they once did, and the springs supplying the famous gardens are drying up. So desperate is the need for charcoal for fuel that the Kashmiri now make it out of tree leaves. This wasteful practice denies the soil of needed organic mulch.

Throughout India, as is well known, cow dung is burned to cook food. The minerals of the dung are not thereby lost, but the ability of dung to improve soil tilth is. Some of the nitrogen in the dung goes off into the air and does not return to Indian soil. Here we see a classic example of the "vicious circle": because Indians are poor they burn dung, depriving the soil of nitrogen and making themselves still poorer the following year. If we give them plenty of food, as they cook this food with cow dung they will lower still more the ability of their land to produce food.

Let us look at another example of this counter-productive behavior. Twenty-five years ago western countries brought food and medicine to Nepal. In the summer of 1974 a disastrous flood struck Bangladesh, killing tens of thousands of people, by government admission. (True losses in that part of the world are always greater than admitted losses.) Was there any connection between feeding Nepal and flooding Bangladesh? Indeed there was, and is.[14]

Nepal nestles amongst the Himalayas. Much of its land is precipitous, and winters are cold. The Nepalese need fuel, which they get from trees. Because more Nepalese are being kept alive now, the demand for timber is escalating. As trees are cut down, the soil under them is washed down the slopes into the rivers that run through India and Bangladesh. Once the absorptive capacity of forest soil is gone, floods rise faster and to higher maxima. The flood of 1974 covered two-thirds of Bangladesh, twice the area of "normal" floods—which themselves are the consequence of deforestation in preceding centuries.

By bringing food and medicine to Nepal we intended only to save lives. But we can never do merely one thing, and the Nepalese lives we saved created a Nepalese energy-famine. The lives we saved from starvation in Nepal a quarter of a century ago were paid for in our time by lives lost to flooding and its attendant evils in Bangladesh. The saying, "Man does not live by bread alone," takes on new meaning.

Still we have not described what may be the worst consequence of a food-only policy: revolution and civil disorder. Many kind-hearted people who support food aid programs solicit the cooperation of "hard-nosed" doubters by arguing that good nutrition is needed for world peace. Starving people will attack others, they say. Nothing could be further from the truth. The monumental studies of Ancel Keys and others have shown that starving people are completely selfish.[15] They are incapable of cooperating with others; and they are incapable of laying plans for tomorrow and carrying them out. Moreover, modern war is so expensive that even the richest countries can hardly afford it.

The thought that starving people can forcefully wrest subsistence from their richer brothers may appeal to our sense of justice, *but it just ain't so.* Starving people fight only among themselves, and that inefficiently.

So what would happen if we brought ample supplies of food to a population that was still poor in everything else? They would still be incapable of waging war at a distance, but their ability to fight among themselves would be vastly increased. With vigorous, well-nourished bodies and a keen sense of their impoverishment in other things, they would no doubt soon create massive disorder in their own land. Of course, they might create a strong and united country, but what is the probability of that? Remember how much trouble the thirteen colonies had in forming themselves into a United States. Then remember that India is divided by two major religions, many castes, fourteen major languages and a hundred dialects. A partial separation of peoples along religious lines in 1947, at the time of the formation of Pakistan and of independent India, cost untold millions of lives. The budding off of Bangladesh (formerly East Pakistan) from the rest of Pakistan in 1971 cost several million more. All these losses were achieved on a low level of nutrition. The possibilities of blood-letting in a population of 600 million well-nourished people of many languages and religions and no appreciable tradition of cooperation stagger the imagination. Philanthropists with any imagination at all should be stunned by the thought of 600 million well-fed Indians seeking to meet their energy needs from their own resources.

So the answer to our Jacobian question, "How can we harm India?" is clear: send food *only*. Escaping the Jacobian by reinverting the question we now ask, "How can we *help* India?" Immediately we see that we must *never* send food without a matching gift of non-food energy. But before we go careening off on an intoxicating new program we had better look at some more quantities.

On a per capita basis, India uses the energy equivalent of one barrel of oil per year; the U.S. uses sixty. The world average of all countries, rich and poor, is ten. If we want to bring India only up to the present world average, we would have to send India about $9 \times 600$ million bbls. of oil per year (or its equivalent in coal, timber, gas or whatever). That would be more than five billion barrels of oil equivalent. What is the chance that we will make such a gift?

Surely it is nearly zero. For scale, note that our total yearly petroleum use is seven billion barrels (of which we import three billion). Of course we use (and have) a great deal of coal too. But these figures should suffice to give a feeling of scale.

More important is the undoubted psychological fact that a fall in income tends to dry up the springs of philanthropy. Despite wide disagreements about the future of energy it is obvious that from now on, for at least the next twenty years and possibly for centuries, our per capita supply of energy is going to fall, year after year. The food we gave in the past was "surplus." By no accounting do we have an energy surplus. In fact, the perceived deficit is rising year by year.

India has about one-third as much land as the United States. She has about three times as much population. If her people-to-land ratio were the same as ours she would have only about seventy million people (instead of 600 million). With the forested and relatively unspoiled farmlands of four centuries ago, seventy million people was probably well within the carrying capacity of the land. Even in today's India, seventy million people could probably make it in comfort and dignity—provided they didn't increase!

To send food only to a country already populated beyond the carrying capacity of its land is to collaborate in the further destruction of the land and the further impoverishment of its people.

Food plus energy is a recommendable policy; but for a large population under today's conditions this policy is defensible only by the logic of the old saying, "If wishes were horses, beggars would ride." The fantastic amount of energy needed for such a program is simply not in view. (We have mentioned nothing of the equally monumental "infrastructure" of political, technological, and educational machinery needed to handle unfamiliar forms and quantities of energy in the poor countries. In a short span of time this infrastructure is as difficult to bring into being as is an abundant supply of energy.)

In summary, then, here are the major foreign-aid possibilities that tender minds are willing to entertain:

a. Food plus energy—a conceivable, but practically impossible program.

b. Food alone—a conceivable and possible program, but one which would destroy the recipient.

In the light of this analysis the question of triage[8] shrinks to negligible importance. If *any* gift of food to overpopulated countries does more harm than good, it is not necessary to decide which countries get the gift and which do not. For posterity's sake we should never send food to any population that is beyond the realistic carrying capacity of its land. The question of triage does not even arise.

Joseph Fletcher neatly summarized this point when he said, "We should give if it helps but not if it hurts." We would do well to memorize his aphorism, but we must be sure we understand the proper object of the verb, which is the recipient. Students of charity have long recognized that an important motive of the giver is to help himself, the giver.[16] Hindus give to secure a better life in the next incarnation; Moslems, to achieve a richer paradise at the end of this life; and Christians in a simpler day no doubt hoped to shorten their stay in purgatory by their generosity. Is there anyone who would say that contemporary charity is completely free of the self-serving element?

To deserve the name, charity surely must justify itself primarily, perhaps even solely, by the good it does the recipient, not only in the moment of giving but in the long run. That every act has multiple consequences was recognized by William L. Davison, who grouped the consequences of an act of charity into two value-classes, positive and negative.[17] True charity, he said,

> confers benefits, and it refrains from injuring. . . . Hence, charity may sometimes assume an austere and even apparently unsympathetic aspect toward its object. When that object's real good cannot be achieved without inflicting pain and suffering, charity does not shrink from the infliction. . . . Moreover, a sharp distinction must be drawn between charity and amiability or good nature—the latter of which is a weakness and may be detrimental to true charity, although it may also be turned to account in its service.

To the ecologically minded student of ethics, most traditional ethics looks like mere amiability, focusing as it does on the manifest misery of the present generation to the neglect of the more subtle but equally real needs of a much larger posterity. It is amiability that feeds the Nepalese in one generation and drowns Bangladeshi in another. It is amiability that, contemplating the wretched multitudes of Indians asks, "How can we let them starve?" implying that we, and only we, have the power to end their suffering. Such an assumption surely springs from hubris.

Fifty years ago India and China were equally miserable, and their future prospects equally bleak. During the past generation we have given India "help" on a massive scale; China, because of political differences between her and us, has received no "help" from us and precious little from anybody else. Yet who is better off today? And whose future prospects look brighter? Even after generously discounting the reports of the first starry-eyed Americans to enter China in recent years, it is apparent that China's 900 million are physically better off than India's 600 million.

All that has come about without an iota of "help" from us.

Could it be that a country that is treated as a responsible agent does better in the long run than one that is treated as an irresponsible parasite which we must "save" repeatedly? Is it not possible that robust responsibility is a virtue among nations as it is among individuals? Can we tolerate a charity that destroys responsibility?

Admittedly, China did not reach her present position of relative prosperity without great suffering, great loss of life. Did millions die? Tens of millions? We don't know. If we had enjoyed cordial relations with the new China during the birth process no doubt we would, out of a rich store of amiability, have seen to it that China remained as irresponsible and miserable as India. Our day-to-day decisions, with their delayed devastation, would have been completely justified by our traditional, posterity-blind ethics which seems incapable of asking the crucial question, *"And then what?"*

Underlying most ethical thought at present is the assumption that human life is the *summum bonum.* Perhaps it is; but we need to inquire carefully into what we mean by "human life." Do we mean the life of each and every human being now living, all 4,000,000,000 of them? Is each presently existing human being to be kept alive (and breeding) regardless of the consequences for future human beings? So, apparently, say amiable, individualistic, present-oriented, future-blind Western ethicists.

An ecologically oriented ethicist asks, "And then what?" and insists that the needs of posterity be given a weighting commensurate with those of the present generation. The economic prejudice that leads to a heavy discounting of the future must be balanced by a recognition that the population of posterity vastly exceeds the population of the living.[18] We know from experience that the environment can be irreversibly damaged and the carrying capacity of a land permanently lowered.[14] Even a little lowering multiplied by an almost limitless posterity should weigh heavily in the scales against the needs of those living, once our charity expands beyond the limits of simple amiability.

We can, of course, increase carrying capacity somewhat. But only hubris leads us to think that our ability to do so is without limit. Despite all our technological accomplishments—and they are many—there is a potent germ of truth in the saying of Horace (65–8 B.C.): *Naturam expelles furca, tamen usque recurret.* "Drive nature off with a pitchfork, nevertheless she will return with a rush." This is the message of Rachel Carson,[19] which has been corroborated by many others.[20]

*The morality of an act is a function of the state of the system at the time the act is performed*—this is the foundation stone of situationist, ecological ethics.[12] A time-blind absolute ethical principle like that implied by the shibboleth, "the sanctity of life," leads to greater suffering than its situationist, ecological alternative—and ultimately and paradoxically, even to a lesser quantity of life over a sufficiently long period of time. The interests of posterity can be brought into the reckoning of ethics if we abandon the idea of the sanctity of (present) life as an absolute ethical ideal, replacing it with the idea of the sanctity of the carrying capacity.

Those who would like to make the theory of ethics wholly rational must look with suspicion on any statement that includes the word "sanctity." There is a whole class of terms whose principal (and perhaps sole) purpose seems to be to set a stop to inquiry: "self-evident" and "sanctity" are members of this class. I must, therefore, show that "sanctity" is used as something more than a discussion-stopper when it occurs in the phrase "the sanctity of the carrying capacity."

Some there are who so love the world of Nature (that is, Nature *sine* Man) that they regard the preservation of a world without humankind as a legitimate objective of human beings. It is difficult to argue this ideal dispassionately and productively. Let me only say that I am not one of this class of nature-lovers; my view is definitely homocentric. Even so I argue that we would do well to accept "Thou shalt not exceed the carrying capacity of any environment" as a legitimate member of a new Decalogue. When for the sake of momentary gain by human beings the carrying capacity is transgressed, the long-term interests of the same human beings—"same" meaning themselves and their successors in time—are damaged. I should not say that the carrying capacity is something that is *intrinsically* sacred (whatever *that* may mean) but that the rhetorical device "carrying capacity" is a shorthand way of dealing time and posterity into the game. A mathematician would, I imagine, view "carrying capacity" as an algorithm, a substitute conceptual element with a different grammar from the elements it replaces. Algorithmic substitutions are made to facilitate analysis; when they are well chosen, they introduce no appreciable errors. I think "carrying capacity" meets significant analytical demands of a posterity-oriented ethics.

In an uncrowded world there may be no ethical need for the ecological concept of the carrying capacity. But ours is a crowded world. We need this concept if we are to minimize human suffering in the long run (and not such a very long run at that). How Western man has pretty well succeeded in locking himself into a suicidal course of action by developing and clinging to a concept of the absolute sanctity of life is a topic that calls for deep inquiry. Lacking the certain knowledge that might come out of such a scholarly investigation, I close this essay with a personal view of the significance of—

## CARRYING CAPACITY*
### (To Paul Sears)

*A man said to the universe:*
*"Sir, I exist!"*
*"However," replied the universe,*
*"The fact has not created in me*
*A sense of obligation."*
    *—Stephen Crane, 1899.*

So spoke the poet, at century's end;
And in those dour days when schools displayed the world,
"Warts and all," to their reluctant learners,
These lines thrust through the layers of wishfulness,
Forming the minds that later found them to be true.

All that is past, now.
Original sin, then mere personal ego,
Open to the shafts of consciousness,
Now flourishes as an ego of the tribe
Whose battle cry (which none dare question) is
"Justice!"—But hear the poet's shade:

A tribe said to the universe,
"Sir, We exist!"
"So I see," said the universe,
"But your multitude creates in me
No feeling of obligation.

"Need creates right, you say? Your need, your right?
Have you forgot we're married?
Humanity and universe—Holy, indissoluble pair!
Nothing you can do escapes my vigilant response.

"Dam my rivers and I'll salt your crops;
Cut my trees and I'll flood your plains.
Kill 'pests' and, by God, you'll get a silent spring!

---
*Copyright: Garrett Hardin, 1975.

Go ahead—save every last baby's life!
I'll starve the lot of them later,
When they can savor to the full
The exquisite justice of truth's retribution.
Wrench from my earth those exponential powers
No wobbling Willie should e'er be trusted with:
Do this, and a million masks of envy shall create
A hell of blackmail and tribal wars
From which civilization will never recover.

"Don't speak to me of shortage. My world is vast
And has more than enough—for no more than enough.
There is a shortage of nothing, save will and wisdom;
But there is a longage of people.

Hubris—that was the Greeks' word for what ails you.
Pride fueled the pyres of tragedy
Which died (some say) with Shakespeare.
O, incredible delusion! That potency should have no limits!
'We believe no evil 'til the evil's done'—
Witness the deserts' march across the earth,
Spawned and nourished by men who whine, 'Abnormal weather.'
Nearly as absurd as crying, 'Abnormal universe!' . . .
But I suppose you'll be saying that, next."

Ravish capacity: reap consequences.
Man claims the first a duty and calls what follows
Tragedy.
Insult—Backlash. Not even the universe can break
This primal link. Who, then, has the power
To put an end to tragedy? Only those who recognize
Hubris in themselves.

## Notes

1. Garrett Hardin, 1968: "The Tragedy of the Commons," *Science*, 162:1243–48.
2. Karl Marx, 1875: "Critique of the Gotha program." (Reprinted in *The Marx-Engels Reader*, Robert C. Tucker, editor. New York: Norton, 1972).
3. Garrett Hardin and John Baden (in press). *Managing the Commons*.

4. Thomas Robert Malthus, 1798: *An Essay on the Principle of Population, as it affects the Future Improvement of Society*. (Reprinted, *inter alia*, by the University of Michigan Press, 1959, and The Modern Library, 1960).

5. Donella H. Meadows, Dennis L. Meadows, Jorgen Randers, and William H. Behrens, 1972: *The Limits to Growth* (New York: Universe Books).

6. C. P. Snow, 1963: *The Two Cultures; and a Second Look* (New York: Mentor).

7. Hans Jonas, 1973: "Technology and Responsibility: Reflections on the New Task of Ethics," *Social Research*, 40:31–54.

8. William and Paul Paddock, 1967: *Famine—1975!* (Boston: Little, Brown & Co.).

9. Garrett Hardin, 1975: "Gregg's Law," *BioScience*, 25:415.

10. Joseph Fletcher, 1966: *Situation Ethics* (Philadelphia: Westminster Press).

11. Helmut Schoeck, 1969: *Envy* (New York: Harcourt, Brace & World).

12. Garrett Hardin, 1972: *Exploring New Ethics for Survival* (New York: Viking).

13. Nicholas Wade, 1974: "Sahelian Drought: No Victory for Western Aid," *Science*, 185:234–37.

14. Erik P. Eckholm, 1975: "The Deterioration of Mountain Environments," *Science*, 189:764–70.

15. Ancel Keys, et al., 1950: *The Biology of Human Starvation*. 2 vols. (Minneapolis: University of Minnesota Press).

16. A. S. Geden, 1928: "Hindu charity (almsgiving)," *Encyclopaedia of Religion and Ethics*, Vol. III, pp. 387–89 (New York: Scribner's).

17. William L. Davidson, 1928: "Charity," *Encyclopaedia of Religion and Ethics*, Vol. III, p. 373 (New York: Scribner's).

18. Garrett Hardin, 1974: "The Rational Foundation of Conservation," *North American Review*, 259(4):14–17.

19. Rachel Carson, 1962: *Silent Spring* (Boston: Houghton Mifflin).

20. M. Taghi Farvar and John P. Milton, editors, 1969: *The Careless Technology* (Garden City, N.Y.: Natural History Press).

# Golden Rice Is Fool's Gold　　　　　　　　　6

GREENPEACE INTERNATIONAL

*Greenpeace International was founded by David McTaggart, a former construction businessman, who in 1971 defied nuclear testing in the South Pacific by the French government with his vessel* Greenpeace III. *He eventually got the French to stop such testing. In 1979, he brought various national Greenpeace organizations together in Europe under the umbrella group Greenpeace International.*

*Beginning in 1979, Greenpeace International fought to save whales, to stop dumping of nuclear waste in oceans, to block the production of toxic wastes, and to prevent oil and mineral drilling in Antarctica. These campaigns have been characterized by a relentless push to seize the attention of international media, sometimes at some risk of the health of Greenpeace members. Greenpeace activist Pete Wilkinson was especially good at such sensationalist events and today works in a media relations company in England that helps pro-environment groups gain publicity and raise money. In 1991, Greenpeace International started to campaign against genetically modified food and crops and won big victories in Great Britain and Europe.*

*Not everyone in Greenpeace agrees with the following press release by Greenpeace International. An ecologist and cofounder of Greenpeace, Dr. Patrick Moore, who served as president of Greenpeace Canada for nine years and on the board of directors of Greenpeace International for seven years, quit Greenpeace International over its condemnation of genetically modified foods.*

GENETICALLY ENGINEERED "Golden Rice" containing provitamin A will not solve the problem of malnutrition in developing countries according to Greenpeace. The Genetic Engineering (GE) industry claims vitamin A rice could save thousands of children from blindness and millions of malnourished people from vitamin A deficiency (VAD) related

---

From a press release from Greenpeace International (www.greenpeace.org/~geneng/) from February 9, 2001. Reprinted by permission of Greenpeace International.

diseases. But a simple calculation based on the product developers' own figures shows an adult would have to eat at least twelve times the normal intake of 300 grams to get the daily recommended amount of provitamin A.

Syngenta, one of the world's leading genetic engineering companies and pesticide producers, which owns many patents on the "Golden Rice", claims a single month of marketing delay of "Golden Rice" would cause 50,000 children to go blind.

Greenpeace calculations show, however, that an adult would have to eat at least 3.7 kilos of dry weight rice, i.e. around 9 kilos of cooked rice, to satisfy his/her daily need of vitamin A from "Golden Rice". In other words, a normal daily intake of 300 grams of rice would, at best, provide 8 percent of the vitamin A needed daily. A breast-feeding woman would have to eat at least 6.3 kilos in dry weight, converting to nearly 18 kilos of cooked rice per day.

"It is clear from these calculations that the GE industry is making false promises about "Golden Rice". "It is a nonsense to think anyone would or could eat this much rice, and there is still no proof that it can provide any significant vitamin benefits anyway," said Greenpeace Campaigner Von Hernandez in the Philippines, where the first grains of the genetically engineered rice had been delivered to the International Rice Research Institute last month for breeding into local rice varieties. "This whole project is actually based on what can only be characterised as intentional deception. We recalculated their figures again and again, we just could not believe serious scientists and companies would do this."

In addition, one of the main sponsors of "Golden Rice", the Rockefeller Foundation, has told Greenpeace the GE industry has "gone too far" in its promotion of the product. While upholding its principal support for the project, Rockefeller Foundation President Gordon Conway wrote to Greenpeace: "the public relations uses of Golden Rice have gone too far. The industry's advertisements and the media in general seem to forget that it is a research product that needs considerable further development before it will be available to farmers and consumers."

"The European markets have resoundingly rejected GE products, consumers worldwide don't want them in their food, and the industry is desperate for alternative markets. "Golden Rice" has been presented as a quick fix for a starving world, but it is swamping attempts to enforce existing effective solutions, and carry out further work on other sustainable, reliable methods to address the problem," added Hernandez.

Genetically engineered rice does not address the underlying causes of vitamin A deficiency (VAD), which are mainly poverty and lack of access to a more diverse diet. For the short term, measures such as supplementation (i.e. pills) and food fortification are cheap and effective. Promoting the use and the access to food naturally rich in provitamin A, such as red palm oil, will also help in addressing the VAD related sufferings. The only long-term solution is to work on the root causes of poverty and to ensure access to a diverse and healthy diet.

# Are We Going Mad? 7

NORMAN BORLAUG

*One episode of the popular TV show* The West Wing *used Norman Borlaug's achievements to illustrate an episode about the need for developed nations to provide technology to the developing world. In this episode, the fictional leader of an African nation makes a brief speech praising Borlaug: "The people who make miracles in the world, one of them lives right here in the United States." Yes, indeed.*

*For nearly sixty years, Borlaug has devoted his life to improving food and food production around the world. Thirty years ago, as director of the Rockefeller Foundation's Mexican wheat program, he developed dwarf wheat, which diverted energy from the stalk into the grain, thus increasing yields. He went on to carry the Green Revolution to Asian nations, such as India and Pakistan, improving wheat and other grains, fighting constant criticism from doomsayers such as Garrett Hardin. At age eighty-four, Borlaug remains an active warrior in the battle to feed and improve the lives of starving millions, especially in sub-Saharan Africa.*

*As the following essay shows, Borlaug is a staunch supporter of biotechnology, and he has no qualms about sending genetically modified food to starving people or using the latest techniques in genetic engineering to grow food for the starving. The essay originally appeared in London's* Independent *newspaper as an open letter to the editor on April 20, 2000.*

April 10, 2000

Open Letter to the Editor
*The Independent* newspaper
London, UK

Dear Sir,

My name is Norman Borlaug. For the past 56 years I have worked as an agricultural scientist in the low-income, food-deficit nations to raise agricul-

---

From Norman Borlaug, "Open Letter to the Editor," London *Independent*, April 20, 2000. Reprinted by permission of the *Independent*.

tural production and productivity. During a recent field tour to Nigeria, Malawi, and Mozambique, I came across an article in The Independent, Thursday, March 30, 2000 issue, titled "America finds ready market for GM Food—the hungry," written by Declan Walsh from Nairobi. A ghastly photograph accompanies the article, depicting a man near death from starvation, lying next to food sacks, with the caption, "A Sudanese man collapsing as he waits for food from the UN World Food Programme. Much of the food donated is genetically modified."

Mr. Walsh's article is seriously biased and misguided. It implies a conspiracy between the U.S. government and the World Food Programme (WFP) to dump unsafe, American genetically modified crops into the one remaining "unquestioning market—emergency aid for the world's starving and displaced." This is unfair.

Mr. Walsh quotes several critics on the use of genetically modified food in Africa. Elfrieda Pschorn-Strauss, from the South African organization Biowatch says, "The US does not need to grow nor donate GM crops. To donate untested food and seed to Africa is not an act of kindness but an attempt to lure Africa into further dependence on foreign aid," she said. Dr. Tewolde Gebre Egziabher of Ethiopia states that "Countries in the grip of a crisis are unlikely to have leverage to say, 'This crop is contaminated; we're not taking it.' They should not be faced with a dilemma between allowing a million people to starve to death and allowing their genetic pool to be polluted." Neither of these individuals offers any credible scientific evidence to back their extreme assertions—which are patently false, in my opinion.

All told, some 400 million tonnes of food are currently traded in world markets. However, only a dozen nations are major food-exporting countries. Among them, the U.S. is by far the largest (80 million tonnes), followed by Argentina, France, Canada, and Australia. The vast majority of this food is sold at market prices around the world, and especially in the industrialized and newly industrializing countries. Still, from a global perspective, without the very large U.S. food surpluses, it is likely that millions would go hungry and even be threatened by starvation.

A relatively small portion of the global food trade goes to emergency food aid organizations like WFP to help feed displaced and hungry people. To my way of thinking, it is fitting that the United States, the world's largest food exporter and wealthiest nation, should be the largest contributor to the WFP, which is responsible for 40–45% of all emergency food aid shipments. Of greater worry to Mr. Walsh should be the fact that international support for emergency food aid has steadily declined in recent years, from about 10 million tonnes in 1994 and 1995, to about 7.7 million tonnes in 1998. Wheat and wheat flour now account for more than half of global food aid. (Incidentally, Mr. Walsh, there are no GM wheats on the market at present.)

WFP only accepts food donations that fully meet the safety standards in the donor country. Since in the US, GM foods are judged to be safe by the Department of Agriculture, Food and Drug Administration, and Environmental Protection Agency, they are acceptable to the WFP. That the EU has placed a two-year moratorium on GM imports says little, per se, about food safety, but rather more about consumer concerns, largely the result of unsubstantiated "scare-mongering" done by GM opponents.

Let's consider the underlying thrust of Walsh's article—that genetically modified food is unnatural and unsafe. "Genetically modified organisms" (GMOs) and "genetically modified foods" (GMFs) are imprecise terms that refer to the use of transgenic crops, i.e., those grown from seeds that contain the genes of different species. The facts are that genetic modification started long before humankind started breeding. Mother Nature did it, and often in a big way. For example, the wheat crops we rely on for much of our food supply are the result of unusual (but natural) crosses between different species of grasses. Today's bread wheat is the result of the hybridization of three different plant genomes, each containing a set of seven chromosomes, and thus could easily be classified as "transgenic." Maize is another crop that is the product of transgenic hybridization. Indeed, it is hard to see how the modern maize plant evolved from Teosinte and Tripsacum—reputed to be its putative (ancient) parents.

Neolithic man—or much more likely women—domesticated virtually all of our food and livestock species over a relatively short period, 10,000 to 15,000 years ago. Subsequently, several hundred generations of farmer-descendents were responsible for making enormous genetic modifications in all of our major crop and animal species. To see how far the evolutionary changes have come, one only needed to look at the 5,000-year old fossilized corn cobs found in the caves of Tehuacan in Mexico, which are about $1/10$ the size of modern maize varieties.

Thanks to the development of science over the past 150 years, we now have the insights into plant genetics and breeding to do purposefully what Mother Nature did herself in the past by chance or design. Genetic modification of crops is not some kind of witchcraft, rather it is the progressive harnessing of the forces of nature to the benefit of feeding the human race.

Indeed, genetic engineering—plant breeding at the molecular level—is just another step in humankind's deepening scientific journey into living genomes. It is not a replacement to conventional breeding but rather a complementary "research tool" to identify desirable genes (traits) from remotely related taxonomic groups and transfer them more quickly and precisely into high-yielding, high-quality crop species.

To date, there has been no credible scientific evidence to suggest that eating transgenic agricultural products damages human health, or the environment. Virtually all of the scientific debate has been [about] possible damage and the risk factor society is willing to take. Certainly, "zero risk" is unrealistic, and probably unattainable. Scientific advance always involves some risk that unintended outcomes can occur. So far, the most prestigious national academies of science, and now even the Vatican, have come out in support of genetic engineering to improve the quantity, quality, and availability of food supplies. The more important matters of concern by civil societies should be equity issues related to genetic ownership, control, and access to transgenic agricultural products.

The closest parallel in my lifetime to the emotional anti-GMO campaign currently under way would be the period, primarily during the time of Stalin, when T. D. Lysenko dominated agriculture in the former Soviet Union. With his brand of ideologically based pseudo-science, Lysenko had some of the Soviet Union's best plant scientists banished, imprisoned, and even killed, among them, the great plant taxonomist, N. T. Vavilov, who probably starved to death in a Soviet concentration camp. The damage to Soviet agriculture caused by Lysenko and his cohorts was enormous, and contributed directly to the collapse of the Soviet Union.

One of the great challenges facing society in the 21st Century will be a renewal and broadening of scientific education—at all age levels—that keeps pace with the times. Nowhere is it more important for "knowledge" to confront "fear born of ignorance" than in the production of food—still the basic human activity. In particular, we need to close the biological science "knowledge gap" in the affluent societies—now thoroughly urban and removed from any tangible relationship to the land. The needless confrontation of consumers against the use of transgenic crop technology in Europe and elsewhere might have been avoided had more people received a better education about genetic diversity and variation.

Privileged societies have the luxury of adopting a very low-risk position on the GM crops issue, even if this action later turns out to be unnecessary.

But the vast majority of humankind does not have such a luxury, and certainly not the hungry victims of wars, natural disasters, and economic crises that are attended by the WFP. I agree with Mr. Walsh when he speculates that "esoteric arguments about the genetic make-up of a bag of grain mean little to those for whom food aid is a matter of life and death." He should take this thought more deeply to heart.

The World Food Programme is one of the shining stars of the UN system, and Ms Catherine Bertini, WFP Executive Director, is a gifted leader and administrator. First elected in 1992, she was re-elected for a second five-year term in 1997. Under her leadership, WFP has achieved the lowest administrative overheads and smallest percentage of staff posted at its headquarters (in Rome) of any UN agency. During 1999, WFP workers and their collaborators helped to feed 86 million people in 82 countries. They have also helped to make significant economic contributions through their efforts to develop food transport and distribution networks and by buying and shipping food from developing countries wherever feasible. WFP staffers are among the world's unsung heroes, who struggle against the clock and under exceedingly difficult conditions to save people from famine. Their achievements, dedication and bravery deserve our highest respect and praise.

Most African nations had the misfortune of gaining independence in the 1960s during the height of the Cold War. They were pulled back and forth between East and West, over-investing in their militaries and under-investing in rural infrastructure, schools and clinics. I fear that Africa, once again, is becoming caught in the crossfire between conflicting economic ideologies and trading blocks—this time centered around science and technology. I am alarmed that some anti-technology elitists are seeking to deny small-scale farmers in sub-Saharan Africa access to improved seeds, fertilizers, and when needed, crop protection chemicals, while they have had the luxury of plentiful and inexpensive food supplies which, in turn, has accelerated economic development. Africa must not be bypassed again!

The world has the agricultural technology—either available or well-advanced in the research pipeline—to feed those 8.3 billion people anticipated in the next quarter of a century. The more pertinent question today is whether farmers and ranchers will be permitted to use that technology. Indeed, extremists in the environmental movement, largely from rich nations and/or the privileged strata of society in poor nations, seem to be doing everything they can to stop scientific progress in its tracks. To be sure, very serious equity barriers exist in the access of the poor to food, and these issues must be addressed and corrected by the world community.

We cannot turn back the clock on agriculture and only use methods that were developed to feed a much smaller population. It took some 10,000 years to expand food production to the current level of about 5 billion tonnes per year. By 2025, we will have to nearly double current production again. This cannot be done unless farmers across the world have access to current high-yielding crop-production methods as well as new biotechnological breakthroughs that can increase the yields, dependability, and nutritional quality of our basic food crops. We need to bring common sense into the debate on agricultural science and technology, and the sooner the better!

Norman E. Borlaug
1970 Nobel Peace Prize

# The Unholy Alliance 8

MAE-WAN HO

*Mae-Wan Ho is one of the world's leading critics of genetically modified food and global industrialized food production. She lectures frequently around the world and frequently attacks industrialized food in her many articles and books.*

*Ho earned a B.S. in biology in 1964 and a Ph.D. in biochemistry in 1967 from Hong Kong University. She was a postdoctoral fellow in biochemical genetics at the University of California in San Diego from 1968 to 1972. During that time, she won a fellowship from the U.S. National Genetics Foundation, which took her to London University, where she became senior research fellow at Queen Elizabeth College. She then became lecturer in genetics in 1976 (equivalent to assistant professor in American universities) and reader (associate professor) in biology in 1985 at the Open University. Ho retired in early June 2000. Since 1994, she has been the scientific advisor to the Third World Network and other public-interest organizations (such as Greenpeace) on biotechnology and biosafety.*

*Ho is the author of* Genetic Engineering: Dream or Nightmare? *The following essay explains why she distrusts genetically modified food in particular and industrialized global food production in general.*

SUDDENLY, the brave new world dawns.

Suddenly, as 1997 begins and the millennium is drawing to a close, men and women in the street are waking up to the realization that genetic engineering biotechnology is taking over every aspect of their daily lives. They are caught unprepared for the avalanche of products arriving, or soon to arrive, in their supermarkets: rapeseed oil, soybean, maize, sugar beet, squash, cucumber . . . It started as a mere trickle less than three years ago—the BST—milk from cows fed genetically engineered bovine growth hormone to boost milk yield, and the tomato genetically engineered to prolong shelf-life. They had provoked so much debate and opposition; as did indeed, the genetic

---

From Mae-Wan Ho, "The Unholy Alliance," *The Ecologist* 27, no. 4 (July/August 1997). Reprinted by permission of *The Ecologist* (www.theecologist.org).

screening tests for an increasing number of diseases. Surely, we wouldn't, and shouldn't, be rushed headlong into the brave new world.

Back then, in order to quell our anxiety, a series of highly publicized "consensus conferences" and "public consultations" were carried out. Committees were set up by many European governments to consider the risks and the ethics, and the debates continued. The public were, however, only dimly aware of critics who deplored "tampering with nature" and "scrambling the genetic code of species" by introducing human genes into animals, and animal genes into vegetables. Warnings of unexpected effects on agriculture and biodiversity, of the dangers of irreversible "genetic pollution", warnings of genetic discrimination and the return of eugenics, as genetic screening and prenatal diagnosis became widely available, were marginalized. So too were condemnations of the immorality of the "patents on life"—transgenic animals, plants and seeds, taken freely by geneticists of developed countries from the Third World, as well as human genes and human cell lines from indigenous peoples.

By and large, the public were lulled into a false sense of security, in the belief that the best scientists and the new breed of "bioethicists" in the country were busy considering the risks associated with the new biotechnology and the ethical issues raised. Simultaneously, glossy information pamphlets and reports, which aimed at promoting "public understanding" of genetic "modification," were widely distributed by the biotech industries and their friends, and endorsed by government scientists. "Genetic modification", we are told, is simply the latest in a "seamless" continuum of biotechnology practised by human beings since the dawn of civilization, from bread and wine-making, to selective breeding. The significant advantage of genetic modification is that it is much more "precise", as genes can be individually isolated and transferred as desired.

Thus, the possible benefits promised to humankind are limitless. There is something to satisfy everyone. For those morally concerned about inequality and human suffering, it promises to feed the hungry with genetically modified crops able to resist pests and diseases and to increase yields. For those who despair of the present global environmental deterioration, it promises to modify strains of bacteria and higher plants that can degrade toxic wastes or mop up heavy metals (contaminants). For those hankering after sustainable agriculture, it promises to develop Greener, more environmentally friendly transgenic crops that will reduce the use of pesticides, herbicides and fertilizers.

That is not all. It is in the realm of human genetics that the real revolution will be wrought. Plans to uncover the entire genetic blueprint of the human being would, we are told, eventually enable geneticists to diagnose, in advance, all the diseases that an individual will suffer in his or her lifetime, even before the individual is born, or even as the egg is fertilized *in vitro*. A whole gamut of specific drugs tailored to individual genetic needs can be designed to cure all diseases. The possibility of immortality is dangling from the horizons as the "longevity gene" is isolated.

There are problems, of course, as there would be in any technology. The ethical issues have to be decided by the public. (By implication, the science is separate and not open to question.) The risks will be minimized. (Again, by implication, the risks have nothing to do with the science.) After all, nothing in life is without risk. Crossing roads is a risk. The new biotechnology (i.e. genetic engineering biotechnology) is under very strict government regulation, and the government's scientists and other experts will see to it that neither the consumer nor the environment will be unduly harmed.

Then came the relaxation of regulation on genetically modified products, on grounds that over-regulation is compromising the "competitiveness" of the industry, and that hundreds of field trials have demonstrated the new biotechnology to be safe. And, in any case, there is no essential difference between transgenic plants produced by the new biotechnology and those produced by conventional breeding methods. (One prominent spokesperson for the industry even went as far as to refer to the varieties produced by conventional breeding methods, *retrospectively*, as "transgenics".[1] This was followed, a year later, by the avalanche of products approved, or seeking, approval marketing, for which neither segregation from non-genetically engineered produce nor labelling is required. One is left to wonder why, if the products are as safe and wonderful as claimed, they could not be segregated, as organic produce has been for years, so that consumers are given the choice of buying what they want.

A few days later, as though acting on cue, the Association of British Insurers announced that, in future, people applying for life policies will have to divulge the results of any genetic tests they have taken. This is seen, by many, as a definite move towards open genetic discrimination. A few days later, a scientist of the Roslin Institute near Edinburgh announced that they had successfully "cloned" a sheep from a cell taken from the mammary gland of an adult animal. "Dolly", the cloned lamb, is now seven months old. Of course it took nearly 300 trials to get one success, but no mention is made of the vast majority of the embryos that failed. Is that ethical? If it can be done on sheep, does it mean it can be done for human beings? Are we nearer to cloning human beings? The popular media went wild with heroic enthusiasm at one extreme to the horror of Frankenstein at the other. Why is this work only coming to public attention now, when the research has actually been going on for at least 10 years?[2]

The public are totally unprepared. They are being plunged headlong, against their will, into the brave new genetically engineered world, in which giant, faceless multinational corporations will control every aspect of their lives, from the food they can eat, to the baby they can conceive and give birth to.

> I should, right away, dispel the myth that genetic engineering is just like conventional breeding techniques. It is not. Genetic engineering by-passes conventional breeding by using the artificially constructed vectors to multiply copies of genes, and in many cases, to carry and smuggle genes into cells. Once inside cells, these vectors slot themselves into the host genome. In this way, transgenic organisms are made carrying the desired transgenes. The insertion of foreign genes into the host genome has long been known to have many harmful and fatal effects including cancer; and this is borne out by the low success rate of creating desired transgenic organisms. Typically, a large number of eggs or embryos have to be injected or infected with the vector to obtain a few organisms that successfully express the transgene.
>
> The most common vectors used in genetic engineering biotechnology are a chimaeric recombination of natural genetic parasites from different sources, including viruses causing cancers and other diseases in animals and plants, with their pathogenic functions "crippled", and tagged with one or more antibiotic resistance "marker" genes, so that cells transformed with the vector can be selected. For example, the vector most widely used in plant genetic engineering is derived from a tumour-inducing plasmid carried by the soil bacterium Agrobacterium tumefaciens. In animals, vectors are constructed from retroviruses causing cancers and other diseases. A vector currently used in fish has a framework from the Moloney marine leukaemic virus, which causes leukaemia in mice, but can infect all mammalian cells. It has bits from the Rous Sarcoma virus, causing sarcomas in chickens, and from the vesicular stomatitis virus, causing oral lesions in cattle, horses, pigs and humans. Such mosaic vectors are particularly hazardous. Unlike natural parasitic genetic elements which have various degrees of host specificity, vectors used in genetic engineering, partly by design, and partly on ac-

count of their mosaic character, have the ability to overcome species barriers, and to infect a wide range of species. Another obstacle to genetic engineering is that all organisms and cells have natural defence mechanisms that enable them to destroy or inactivate foreign genes, and transgene instability is a big problem for the industry. Vectors are now increasingly constructed to overcome those mechanisms that maintain the integrity of species. The result is that the artificially constructed vectors are especially good at carrying out horizontal gene transfer.

Let me summarize why rDNA technology differs radically from conventional breeding techniques.

- I. Genetic engineering recombines genetic material in the laboratory between species that do not interbreed in nature.
- 2. While conventional breeding methods shuffle different forms (alletes) of the same genes, genetic engineering enables completely new (exotic) genes to be introduced with unpredictable effects on the physiology and biochemistry of the resultant transgenic organism.
- 3. Gene multiplications and a high proportion of gene transfers are mediated by vectors which have the following undesirable characteristics:
- a. many are derived from disease-causing viruses, plasmids and mobile genetic elements—parasitic DNA that have the ability to invade cells and insert themselves into the cell's genome causing genetic damages.
- b. they are designed to break down species barriers so that they can shuttle genes between a wide range of species. Their wide host range means that they can infect many animals and plants, and in the process pick up genes from viruses of all these species to create new pathogens.
- c. they routinely carry genes for antibiotic resistance, which is already a big health problem.
- d. they are increasingly constructed to overcome the recipient species' defence mechanisms that break down or inactivate foreign DNA.

Isn't it a bit late in the day to tell us that? you ask. Yes and no. Yes, because I, who should, perhaps, have known better, was caught unprepared like the rest. And no, because there have been so many people warning us of that eventuality, who have campaigned tirelessly on our behalf, some of them going back to the earliest days of genetic engineering in the 1970s—although we have paid them little heed. No, it is not too late, if only because that is precisely what we tend to believe, and are encouraged to believe. A certain climate is created—that of being rapidly overtaken by events—reinforcing the feeling that the tidal wave of progress brought on by the new biotechnology is impossible to stem, so that we may be paralysed into accepting the inevitable. No, because we shall not give up, for the consequence of giving up is the brave new world, and soon after that, there may be no world at all. The gene genie is fast getting out of control. The practitioners of genetic engineering biotechnology, the regulators and the critics alike, have *all* underestimated the risks involved, which are *inherent* to genetic engineering biotechnology, particularly as misguided by an outmoded and erroneous world-view that comes from bad science. The dreams may already be turning into nightmares.

That is why people like myself are calling for an immediate moratorium on further releases and marketing of genetically engineered products, and for an independent public enquiry to be set up to look into the risks and hazards involved, taking into account the most comprehensive, scientific knowledge in addition to the social, moral implications. This would be most timely, as public opposition to genetic engineering biotechnology has been gaining momentum throughout Europe and the USA.

In Austria, a record 1.2 million citizens, representing 20 per cent of the electorate, have signed a people's petition to ban genetically engineered foods, as well as deliberate releases of genetically modified organisms and patenting of life. Genetically modified foods were also rejected earlier by a lay people consultation in Norway, and by 95 per cent of consumers in Germany, as revealed by a recent survey. The European Parliament has voted by an overwhelming 407 to 2 majority to censure the Commission's authorization, in December 1996, for imports of Ciba-Geigy's transgenic maize into Europe, and is calling for imports to be suspended while the authorization is re-examined. The European Commission has decided that in the future genetically engineered seeds will be labelled, and is also considering proposals for retroactive labelling. Commissioner Emma Bonino is to set up a new scientific committee to deal with genetically engineered foods, members of which are to be completely independent of the food industry. Meanwhile, Franz Fischler, the European Commissioner on Agriculture, supports a complete segregation and labelling of production lines of genetically modified and non-genetically modified foods.

In June this year, President Clinton imposed a five-year ban on human cloning in the USA [this ban applied only to research using federal funds—*Ed.*], while the UK House of Commons Science and Technology Committee (STC) wants British law to be amended to ensure that human cloning is illegal. The STC, President Chirac of France and German Research Minister Juergen Ruettgers are also calling for an international ban on human cloning.

Like other excellent critics before me,[3] I do not think there is a grand conspiracy afoot, though there are many forces converging to a single terrible end. Susan George comments, "They don't have to conspire if they have the same world-view, aspire to similar goals and take concerted steps to attain them."[4]

I am one of those scientists who have long been highly critical of the reductionist mainstream scientific world-view, and have begun to work towards a radically different approach for understanding nature.[5] But I was unable, for a long time, to see how much science really matters in the affairs of the real world, not just in terms of practical inventions like genetic engineering, but in how that scientific world-view takes hold of people's unconscious, so that they take action, involuntarily, unquestioningly, to shape the world to the detriment of human beings. I was so little aware of how that science is used, without conscious intent, to intimidate and control, to obfuscate, to exploit and oppress; how that dominant world-view generates a selective blindness to make scientists themselves ignore or misread scientific evidence.

The point, however, is not that *science* is bad—but that there can be *bad science* that ill-serves humanity. Science can often be wrong. The history of science can just as well be written in terms of the mistakes made than as the series of triumphs it is usually made out to be. Science is nothing more, and nothing less, than a system of concepts for understanding nature and for obtaining reliable knowledge that enables us to live sustainably with nature. In that sense, one can ill-afford to give up science, for it is through our proper understanding and knowledge of nature that we can live a satisfying life, that we can ultimately distinguish the good science, which serves humanity, from the bad science that does not. In this view, science is imbued with moral values from the start, and cannot be disentangled from them. Therefore it is bad science that purports to be "neutral" and divorced from moral values, as much as it is bad science that ignores scientific evidence.

It is clear that I part company with perhaps a majority of my scientist colleagues in the mainstream, who believe that science can never be wrong, although it can be misused. Or else they carefully distinguish science, as neutral and value-free, from its application, technology, which can do harm or good.[6] This distinction between science and technology is spurious, especially in the case of an experimental science like genetics, and almost all of biology, where the techniques determine what sorts of question are asked and hence the range of answers that are important, significant and relevant to the science. Where would molecular genetics be without the tools that enable practitioners to recombine and manipulate our destiny? It is an irresistibly heroic view, except that it is totally wrong and misguided.

It is also meaningless, therefore, to set up Ethical Committees which do not question the basic scientific assumptions behind the practice of genetic engineering biotechnology. Their brief is severely limited, often verging on the trivial and banal—such as whether a pork gene transferred to food plants might be counter to certain religious beliefs—in comparison with the much more fundamental questions of eugenics, genetic discrimination and, indeed, whether gene transfers should be carried out at all. They can do nothing more than make the unacceptable acceptable to the public.

The debate on genetic engineering biotechnology is dogged by the artificial separation imposed between "pure" science and the issues it gives rise to. "Ethics" is deemed to be socially determined, and therefore negotiable, while the science is seen to be beyond reproach, as it is the "laws" of nature. The sames goes for the distinction between "technology"—the application of science—and the science. Risk assessments are to do with the technology, leaving the science equally untouched. The technology can be bad for your health, but not the science. In this article, I shall show why science cannot be separated from moral values nor from the technology that shapes our society. In other words, bad science is unquestionably bad for one's health and wellbeing, and should be avoided at all costs. Science is, above all, fallible and negotiable, because we have the choice, to do or not to do. It should be negotiated for the public good. That is the only ethical position one can take with regard to science. Otherwise, we are in danger of turning science into the most fundamentalist of religions, that, working hand in hand with corporate interests, will surely usher in the brave new world.

## Bad Science and Big Business

What makes genetic engineering biotechnology dangerous, in the first instance, is that it is an unprecedented, close alliance between two great powers that can make or break the world: science and commerce. Practically all established molecular geneticists have some direct or indirect connection with industry, which will set limits on what the scientists can and will do research on, not to mention the possibility of compromising their integrity as independent scientists.[7]

The worst aspect of the alliance is that it is between the most reductionist science and multinational monopolistic industry at its most aggressive and exploitative. If the truth be told, it is bad science working together with big business for quick profit, aided and abetted by our governments for the banal reason that governments wish to be re-elected to remain in "power".[8]

Speaking as a scientist who loves and believes in science, I have to say it is bad science that has let the world down and caused the major problems we now face, not the least among which is by promoting and legitimizing a particular world-view. It is a reductionist, manipulative and exploitative world-view. Reductionist because it sees the world as bits and pieces, and denies there are organic wholes such as organisms, ecosystems, societies and community of nations. Manipulative and exploitative because it regards nature and fellow human beings as objects to be manipulated and exploited for gain; life being a Darwinian struggle for survival of the fittest.

It is by no means coincidental that the economic theory currently dominating the world is rooted in the same *laissez-faire* capitalist ideology that gave rise to Darwinism. It acknowledges no values other than self-interest, competitiveness and the accumulation of wealth, at which the developed nations have been very successful. Already, according to the 1992 United Nations Development Programme Report, the richest fifth of the world's population has amassed 82.7 per cent of the wealth, while the poorest fifth gets a piddling 1.4 per cent. Or, put in another way, there are now 477 billionaires in the world whose combined assets are roughly equal to the combined annual incomes of the poorer half of humanity—2.8 billion people.[9] Do we need to be more "competitive" still to take from the poorest their remaining pittance? That is, in fact, what we are doing.

The government representatives of the superpowers are pushing for a "globalized economy" under trade agreements which erase all economic borders. "Together, the processes of deregulation and globalization are undermining the power of both unions and governments and placing the power of global corporations and finance beyond the reach of public accountability."[10] The largest corporations continue to consolidate that power through mergers, acquisitions and strategic alliances. Multinational corporations now comprise 51 of the world's 100 largest economies: only 49 of the latter are nations.

By 1993, agricultural biotechnology was being controlled by just 11 giant corporations, and these are now undergoing further mergers. The OECD (Organization for Economic Co-operation and Development) member countries are at this moment working in secret in Paris on the Multilateral Agreements on Investment (MAI), which is written by and for corporations to prohibit any government from establishing performance or accountability standards for foreign investors. European Commissioner, Sir Leon Brittan, is negotiating in the World Trade Organization, on behalf of the European Community, to ensure that no barriers of any kind should remain in the South to dampen exploitation by the North, and at the same time, to protect the deeply unethical "patents of life" through Trade Related Intellectual Property Rights (TRIPS) agreements.[11] So, in addition to gaining complete control of the food supply of the South through exclusive rights to genetically engineered seeds, the big food giants of the North can asset-strip the South's genetic and intellectual resources with impunity, up to and including genes and cell lines of indigenous peoples.

There is no question that the mindset that leads to and validates genetic engineering is *genetic determinism*—the idea that organisms are determined by their genetic makeup, or the totality of their genes. Genetic determinism derives from the marriage of Darwinism and Mendelian genetics. For those imbued with the mindset of genetic determinism, the major problems of the world can be solved simply by identifying and manipulating genes, for genes determine the characters of organisms; so by identifying a gene we can predict a desirable or undesirable trait, by changing a gene we change the trait, by transferring a gene we transfer the corresponding trait.

The Human Genome Project was inspired by the same genetic determinism that locates the "blueprint" for constructing the human being in the human genome. It may have been a brilliant political move to capture research funds and, at the same time, to revive a flagging pharmaceutical industry, but its scientific content was suspect from the first.

Genetic engineering technology promises to work for the benefit of mankind; the reality is something else.

- It displaces and marginalizes all alternative approaches that address the social and environmental causes of malnutrition and ill-health, such as poverty and unemployment, and the need for a sustainable agriculture that could regenerate the environment, guarantee long-term food security and, at the same time, conserve indigenous biodiversity.

- Its purpose is to accommodate problems that reductionist science and industry have created in the first place—widespread environmental deterioration from the intensive, high-input agriculture of the Green Revolution, and accumulation of toxic wastes from chemical industries. What's offered now is more of the same, except with new problems attached.
- It leads to discriminatory and other unethical practices that are against the moral values of societies and community of nations.
- Worst of all, it is pushing a technology that is untried, and, according to existing knowledge, is inherently hazardous to health and biodiversity.

Let me enlarge on that last point here, as I believe it has been underestimated, if not entirely overlooked by the practitioners, regulators and many critics of genetic engineering biotechnology alike, on account of a certain blindness to concrete scientific evidence, largely as a result of their conscious or unconscious commitment to an old, discredited paradigm. The most immediate hazards are likely to be in public health—which has already reached a global crisis, attesting to the failure of decades of reductionist medical practices—although the hazards to biodiversity will not be far behind.

## Genetic Engineering Biotechnology Is Inherently Hazardous

According to the 1996 World Health Organization Report, at least 30 new diseases, including AIDS, Ebola and Hepatitis C, have emerged over the past 20 years, while old infectious diseases such as tuberculosis, cholera, malaria and diphtheria are coming back worldwide. Almost every month now in the UK we hear reports on fresh outbreaks: *Streptococcus*, meningitis, *E. coli*. Practically all the pathogens are resistant to antibiotics, many to multiple antibiotics. Two strains of *E. coli* isolated in a transplant ward outside Cambridge in 1993 were found to be resistant to 21 out of 22 common antibiotics.[12] A strain of *Staphylococcus* isolated in Australia in 1990 was found to be resistant to 31 different drugs.[13] Infections with these and other strains will very soon become totally invulnerable to treatment. In fact, scientists in Japan have already isolated a strain of *Staphylococcus aureus* that is resistant even to the last resort antibiotic, vancomycin.[14]

Geneticists have now linked the emergence of pathogenic bacteria and of antibiotic resistance to *horizontal gene transfer*—the transfer of genes to unrelated species, by infection through viruses, through pieces of genetic material, DNA, taken up into cells from the environment, or by unusual mating taking place between unrelated species. For example, horizontal gene transfer and subsequent genetic recombination have generated the bacterial strains responsible for the cholera outbreak in India in 1992,[15] and the *Streptococcus* epidemic in Tayside in 1993.[16] The *E. coli* 157 strain involved in the recent outbreaks in Scotland is believed to have originated from horizontal gene transfer from the pathogen, *Shigella*.[17] Many unrelated bacterial pathogens, causing diseases from bubonic plague to tree blight, are found to share an entire set of genes for invading cells, which have almost certainly spread by horizontal gene transfer.[18] Similarly, genes for antibiotic resistance have spread horizontally and recombined with one another to generate multiple antibiotic resistance throughout the bacterial populations.[19] Antibiotic resistance genes spread readily by contact between human beings, and from bacteria inhabiting the gut of farm animals to those in human beings.[20] Multiple antibiotic resistant strains of pathogens have been endemic in many hospitals for years.[21]

What is the connection between horizontal gene transfer and genetic engineering? Genetic engineering is a technology designed specifically to transfer genes horizontally between species that do not interbreed. It is designed to break down species barriers and, increasingly, to overcome the species' defence mechanisms which normally degrade or inactivate foreign genes.[22] For the purpose of manipulating, replicating and transferring genes, genetic engineers make use of recombined versions of precisely those genetic parasites causing diseases including cancers, and others that carry and spread virulence genes and antibiotic resistance genes. Thus the technology will contribute to an increase in the frequency of horizontal gene transfer of those genes that are responsible for virulence and antibiotic resistance, and allow them to recombine to generate new pathogens.

What is even more disturbing is that geneticists have now found evidence that the presence of antibiotics typically increases the frequency of horizontal gene transfer 100-fold or more, possibly because the antibiotic acts like a sex hormone for the bacteria, enhancing mating and exchange of genes between unrelated species.[23] Thus, antibiotic resistance and multiple antibiotic resistance cannot be overcome simply by making new antibiotics, *for antibiotics create the very conditions to facilitate the spread of resistance.* The continuing profligate use of antibiotics in intensive farming and in medicine, in combination with the commercial-scale practice of genetic engineering, may already be major contributing factors for the accelerated spread of multiple antibiotic resistance among new and old pathogens that the WHO 1996 Report has identified within the past 10 years. For example, there has been a dramatic rise both in terms of incidence and severity of cases of infections by *Salmonella*,[24] with some countries in Europe witnessing a staggering 20-fold increase in incidence since 1980.

That is not all. One by one, those assumptions on which geneticists and regulatory committees have based their assessment of genetically engineered products to be "safe" have fallen by the wayside, especially in the light of evidence emerging within the past three to four years. However, there is still little indication that the new findings are being taken on board. On the contrary, regulatory bodies have succumbed to pressure from the industry to relax already inadequate regulations. Let me list a few more of the relevant findings in genetics.

We have been told that horizontal gene transfer is confined to bacteria. That is not so. It is now known to involve practically all species of animal, plant and fungus. It is possible for any gene in any species to spread to any other species, especially if the gene is carried on genetically engineered gene-transfer vectors. Transgenes and antibiotic resistance marker genes from transgenic plants have been shown to end up in soil fungi and bacteria.[25] The microbial populations in the environment serve as the gene-transfer highway and reservoir, supporting the replication of the genes and allowing them to spread and recombine with other genes to generate new pathogens.[26]

We have been assured that "crippled" laboratory strains of bacteria and viruses do not survive when released into the environment. That is not true. There is now abundant evidence that they can either survive quite well and multiply, or they can go dormant and reappear after having acquired genes from other bacteria to enable them to multiply.[27] Bacteria co-operate much more than they compete. They share their most valuable assets for survival.

We have been told that DNA is easily broken down in the environment. Not so. DNA can remain in the environment where they can be picked up by bacteria and incorporated into their genome.[28] DNA is, in fact, one of the toughest molecules. Biochemists jumped with joy when they didn't have to work with proteins anymore, which lose their activity very readily. By contrast, DNA survives rigorous boiling, so when they approve processed food on grounds that there can be no DNA left, ask exactly how the processing is done, and whether the appropriate tests for the presence of DNA have been carried out.

The survival of "crippled" laboratory strains of bacteria and viruses and the persistence of DNA in the environment are of particular relevance to the so-called "contained" users producing transgenic pharmaceuticals, enzymes and food additives. "Tolerated" releases and transgenic wastes from such users may already have released large amounts of transgenic bacteria and viruses as well as DNA into the environment since the early 1980s when commercial genetic engineering biotechnology began.

We are told that DNA is easily digested by enzymes in our gut. Not true. The DNA of a virus has been found to survive passage through the gut of mice. Furthermore, the DNA readily finds its way into the bloodstream, and into all kinds of cell in the body.[29] Once inside the cell, the DNA can insert itself into the cell's genome, and create all manner of genetic disturbances, including cancer.[30]

There are yet further findings pointing to the potential hazards of generating new disease-causing viruses by recombination between artificial viral vectors and vaccines and other viruses in the environment. The viruses generated in this way will have increased host ranges, infecting and causing diseases in more than one species, and hence very difficult to eradicate. *We are already seeing such viruses emerging.*

- Monkeypox, a previously rare and potentially fatal virus caught from rodents, is spreading through central Zaire.[31] Between 1981–1986 only 37 cases were known, but there have been at least 163 cases in one eastern province of Zaire alone since July 1995. For the first time, humans are transmitting the disease directly from one to the other.
- An outbreak of hantavirus infection hit southern Argentina in December 1996, the first time the virus was transmitted from person to person.[32] Previously, the virus was spread by breathing in the aerosols from rodent excrement or urine.
- New highly virulent strains of infectious bursal disease virus (IBDV) spread rapidly throughout most of the poultry industry in the Northern Hemisphere, and are now infecting Antarctic penguins, and are suspected of causing mass mortality.[33]
- New strains of distemper and rabies viruses are spilling out from towns and villages to plague some of the world's rarest wild animals in Africa:[34] lions, panthers, wild dogs, giant otter.

None of the plethora of new findings has been taken on board by the regulatory bodies. On the contrary, safety regulations have been relaxed. The public is being used, against its will, as guinea pigs for genetically engineered products, while new viruses and bacterial pathogens may be created by the technology every passing day.

The present situation is reminiscent of the development of nuclear energy which gave us the atom bomb, and the nuclear power stations that we now know to be hazardous to health and also to be environmentally unsustainable on account of the long-lasting radioactive wastes they produce. Joseph Rotblat, the British physicist who won the 1995 Nobel Prize after years of battling against nuclear weapons, has this to say. "My worry is that other advances in science may result in other means of mass destruction, maybe more readily available even than nuclear weapons. Genetic engineering is quite a possible area, because of these dreadful developments that are taking place there."[35]

The large-scale release of transgenic organisms is much worse than nuclear weapons or radioactive nuclear wastes, as genes can replicate indefinitely, spread and recombine. There may yet be time enough to stop the industry's dreams turning into nightmares if we act now, before the critical genetic "melt-down" is reached.

## Notes

[Editor's note: no reference list appeared with this essay in *The Ecologist*.]

1. The first time I heard the word "transgenic" being used on cultivars resulting from conventional breeding methods was from Henry Miller, a prominent advocate for genetic engineering biotechnology, in a public debate with myself, organized by the Oxford Centre for Environment, Ethics and Society, in Oxford University on February 20, 1997.

2. "Scientists scorn sci-fi fears over sheep clone" The Guardian, February 24, 1997, p.7. Lewis Wolpert, development biologist at University College London, was reported as saying, "It's a pretty risky technique with lots of abnormalities." Also report and interview in the Eight O'Clock News, BBC Radio 4, February 24, 1997.

3. As for instance, Spallone, 1992.

4. George, 1998, p.5.

5. My colleague Peter Saunders and I began working on an alternative approach to neo-Darwinian evolutionary theory in the 1970s. Major collections of multi-author essays appeared in Ho and Saunders, 1984; Pollard, 1981; Ho and Fox, 1988.

6. Lewis Wolpert, who currently heads the Committee for the Public Understanding of Science, argues strenuously for this "fundamentalist" view of science. See Wolpert, 1996.

7. See Hubbard and Wald, 1993.

8. This was pointed out to me by Martin Khor, during a course on Globalization and Economics that he gave at Schumacher College, February 3–10, 1997.

9. See Korten, 1997.

10. Korten, 1997, p.2.
11. See Perlas, 1994; also WTO: New Setback for the South, Third World Resurgence issue 77/78, 1997, which contains many articles reporting on the WTO meeting held in December 1996 in Singapore.
12. Brown et al., 1993.
13. Udo and Grubb, 1990.
14. "Superbug spectre haunts Japan", Michael Day, New Scientist 3 May, 1997, p.5.
15. See Bik et al, 1995; Prager et al., 1995; Reidl and Makalanos, 1995.
16. Whatmore et al., 1994; Kapur et al., 1995; Schnitzler et al., 1995; Upton et al., 1996.
17. Professor Hugh Pennington, on BBC Radio 4 News, February 1997.
18. Baringa, 1996.
19. Reviewed by Davies, 1994.
20. Tschape, 1994.
21. See World Health Report, 1996; also Garret, 1995, chapter 13, for an excellent account of the history of antibiotic resistance in pathogens.
22. See Ho and Tappeser, 1997.
23. See Davies, 1994.
24. WHO Fact Sheet No. 139, January 1997.
25. Hoffman et al., 1994; Schluter et al., 1995.
26. See Ho, 1996a.
27. Jager and Tappeser, 1996, have extensively reviewed the literature on the survival of bacteria and DNA released into different environments.
28. See Lorenz and Wackernagel, 1994.
29. See Schubert et al., 1994; also New Scientist January 24, p.24, featured a short report on recent findings of the group that were presented at the International Congress on Cell Biology in San Francisco, December 1996.
30. Wahl et al., 1984; see also relevant entries in Kendrew, 1995, especially "slow transforming retroviruses" and "Transgenic technologies".
31. "Killer virus piles on the misery in Zaire" Debora MacKenzie, New Scientist April 19, 1997, p.12.
32. "Virus gets personal" New Scientist April 26, 1997, p.13.
33. "Poultry virus infection in Antarctic penguins" Heather Gardner, Knowles Kerry and Martin Riddle Nature 387, May 15, 1997, p.245.
34. See Pain, 1997.
35. Quoted in "The spectre of a human clone" The Independent, February 26, 1997, p.1.

# The FDA's Volte-Face on Food Biotech                    9

HENRY I. MILLER

*Henry I. Miller, M.S., M.D., is a senior research fellow at the Hoover Institution, a conservative think tank located on the campus of Stanford University in California. He writes about applied science and technology, especially genetic engineering, and governments' attempts to make such products safe. His work emphasizes the excessive costs of government regulation and models for regulatory reform.*

*Miller joined the Food and Drug Administration (FDA) in 1979 and served in a number of posts involved with new biotechnology. He was the medical reviewer for the first genetically engineered drugs evaluated by the FDA and was instrumental in the rapid licensing of human insulin and human growth hormone. He has been special assistant to the FDA's commissioner, with responsibility for biotechnology issues, and from 1989 to 1994, he was the founding director of the FDA's Office of Biotechnology.*

*In the essay that follows, Miller criticizes his former employer and argues that the FDA is requiring too many regulatory hurdles and too much testing for genetically engineered foods and that the FDA should have stuck by its earlier, more reasonable decisions.*

ONE OF THE FDA's few regulatory successes during the past decade has been its oversight of biotechnology-derived, or gene-spliced, foods. For the past eight years, the agency's official policy has treated gene-spliced and other foods the same, and required scrutiny by regulators only when there are specific safety concerns. This approach was widely applauded as regulation that makes sense, protects consumers, and permits innovation.

However, under pressure from anti-technology extremists and the Clinton administration, the agency this week announced a change in policy that re-

---

From Henry I. Miller, "FDA Plan Will Jeopardize Food Biotech," unpublished manuscript, which appeared on the now-defunct AgBioView listserv (http://agbioview.listbot.com). Reprinted by permission of the author.

verses both its scientific approach to food regulation and a twenty-year old commitment not to discriminate against biotechnology-derived foods and pharmaceuticals.

The new approach is tantamount to singling out only cars with advanced engineering for a punitive tax. The result will be, in the long term, international bureaucracies meddling where they don't belong, disuse of a stunning new technology, diminished choices for farmers and consumers, and higher food prices.

Thousands of biotech foods in US supermarkets have been regulated under the FDA's 1992 policy on products from "new plant varieties," which applied irrespective of whether the plant arose by gene-splicing or conventional genetic engineering methods. It defined certain potentially hazardous characteristics of new foods that, if present, required greater scrutiny by the agency, and which could have resulted in additional testing and labeling, or banishment from commerce. Thus, the agency's approach conformed to the fundamental principle that the degree of scrutiny should be commensurate with risk. Likewise, it was consistent with a widely held scientific consensus that "conventional" and new biotechnology are essentially equivalent, and that the highly precise gene-splicing techniques, in fact, yield a better characterized and more predictable product.

At the same time that the official FDA policy treated biotech foods no differently from others, the agency maintained a "voluntary consultation procedure," in which producers of biotech foods were expected to consult with the agency before marketing their products, and without exception they did so. The major change announced on Wednesday would require the producers to notify the FDA four months before marketing a gene-spliced food and provide the agency with data that affirm the new food's safety.

What's so wrong with codifying essentially what was previously voluntary, but standard, practice? Plenty.

First, the data requirements of the new policy are excessive. The FDA lists nine categories of obligatory information whose level of detail is far greater than would be required (or possible) for food products made with less precise, less sophisticated techniques.

Second, the new policy reverses the FDA's twenty-year-old guiding principles for oversight of biotechnology—that regulation should focus on real risks and should not turn on the use of one technique or another. These tenets have provided effective oversight for thousands of new biotechnology products, including foods, drugs, vaccines and diagnostic tests.

Finally, the FDA's abandonment of a scientific approach to biotech regulation has greater implications, both geopolitical and temporal, as I saw firsthand in March during a meeting in Japan of a task force of the UN's food standards organization, the Codex Alimentarius Commission. The change of policy at the FDA (which was then impending) strikingly altered the dynamics of the negotiation.

The FDA has for decades been considered a world leader in biotech regulation and could be relied upon in international forums to defend scientific principles and vigorously advocate its own risk-based approach. Faced with initial antagonism to the US position from other countries and NGOs, which is not unusual at international negotiations on regulatory issues, the US delegation commonly would set the tone by insisting on adherence to scientific principles and explaining the scientific basis for its own regulatory policy.

What was anomalous at the Codex task force meeting was that the US delegate (and senior FDA food regulator), Robert Lake, never mentioned the FDA's own risk-based approach. He never cited the important principle that the degree of regulatory scrutiny should be commensurate with risk. Nor did he invoke the scientific consensus about the essential equivalence between old and new biotech.

Instead, the US followed the lead of the European Commission and France, both vehemently anti-biotech, and agreed to work toward Draconian and unscientific standards for gene-spliced foods. This first session of the Codex task force, which is scheduled to complete its work in 2003, was dominated by the relentlessly Luddite European Commission, which advocates both the creation of overt obstacles to the use of gene-splicing techniques in food production and agriculture, and also vagueness in regulatory definitions and concepts. This ensures that regulators can be as arbitrary and capricious toward biotech products as they wish. The prospect of unscientific, overly burdensome Codex standards for gene-spliced foods is ominous, because members of the WTO will, in principle, be required to follow them, and they will provide cover for unfair trade practices.

The impending deterioration in domestic regulatory policy—that is, the changes just announced—tied the hands of the US delegation in Japan, and will continue to do so in other international forums that are addressing biotech food regulation. These include two other Codex panels and the Paris-based Organization for Economic Cooperation and Development. International regulation is destined to become biotech food's bête noire.

The long-standing FDA policy toward gene-spliced and other novel foods worked admirably. It involved the government only in those extraordinarily rare instances when products raised safety issues. For others, market forces were permitted to work their magic, the result of which was eight years of unprecedented choice for farmers, food producers and consumers. The policy also encouraged strong FDA advocacy for scientific regulation internationally, which has now ended with dire consequences.

Food production has low profit margins and cannot easily absorb the costs of gratuitous regulation, domestic or international. The overregulation of gene-spliced foods will prevent its wide application to food production, deprive farmers of important tools for raising productivity, and deny to food manufacturers and consumers greater choice among improved, innovative products.

# Dr. Strangelunch
# Why We Should Learn to Love Genetically Modified Food

RONALD BAILEY

*Ronald Bailey is an environmental journalist with a commitment to sound science. He is the science correspondent for* Reason *magazine; the author of several books, including* Eco-Scam: The False Prophets of Ecological Apocalypse; *and the editor of* The True State of the Planet *and* Earth Report 2000: The True State of the Planet. *His articles and reviews have appeared in the* Wall Street Journal, Washington Post, Commentary, New York Times Book Review, The Public Interest, Smithsonian Magazine, National Review, Reason, Forbes, *the* Washington Times, Newsday, *and* Readers Digest.

*Bailey has produced several series and documentaries for PBS television and ABC News. He has lectured at Harvard University, Rutgers University, McGill University, University of Alaska, Université de Quebec à Montreal, the Cato Institute, the Instituto de Libertad y Desarrollo (Chile), and the American Enterprise Institute.*

*In the piece that follows, Bailey argues that opposition by many activists against GM crops and GM food is a new form of Luddism (after the group in Great Britain that wanted to smash the machines of the Industrial Revolution). He also argues that their objections are not really from fears about safety or the environment but from ideological objections to capitalism, international markets, and the slow pace of democratic change.*

TEN THOUSAND PEOPLE were killed and 10 to 15 million left homeless when a cyclone slammed into India's eastern coastal state of Orissa in October 1999. In the aftermath, CARE and the Catholic Relief

---

Reprinted, with permission, from the January 2001 issue of *Reason* magazine. Copyright 2001 by Reason Foundation, 3415 S. Sepulveda Blvd., Suite 400, Los Angeles, CA 90034. www.reason.com.

Society distributed a high-nutrition mixture of corn and soy meal provided by the U.S. Agency for International Development to thousands of hungry storm victims. Oddly, this humanitarian act elicited cries of outrage.

"We call on the government of India and the state government of Orissa to immediately withdraw the corn-soya blend from distribution," said Vandana Shiva, director of the New Delhi–based Research Foundation for Science, Technology, and Ecology. "The U.S. has been using the Orissa victims as guinea pigs for GM [genetically modified] products which have been rejected by consumers in the North, especially Europe." Shiva's organization had sent a sample of the food to a lab in the U.S. for testing to see if it contained any of the genetically improved corn and soy bean varieties grown by tens of thousands of farmers in the United Sates. Not surprisingly, it did.

"Vandana Shiva would rather have her people in India starve than eat bioengineered food," says C. S. Prakash, a professor of plant molecular genetics at Tuskegee University in Alabama. Per Pinstrup-Andersen, director general of the International Food Policy Research Institute, observes: "To accuse the U.S. of sending genetically modified food to Orissa in order to use the people there as guinea pigs is not only wrong; it is stupid. Worse than rhetoric, it's false. After all, the U.S. doesn't need to use Indians as guinea pigs, since millions of Americans have been eating genetically modified food for years now with no ill effects."

Shiva not only opposes the food aid but is also against "golden rice," a crop that could prevent blindness in half a million to 3 million poor children a year and alleviate vitamin A deficiency in some 250 million people in the developing world. By inserting three genes, two from daffodils and one from a bacterium, scientists at the Swiss Federal Institute of Technology created a variety of rice that produces the nutrient beta-carotene, the precursor to vitamin A. Agronomists at the International Rice Research Institute in the Philippines plan to crossbreed the variety, called "golden rice" because of the color produced by the beta-carotene, with well-adapted local varieties and distribute the resulting plants to farmers all over the developing world.

Last June, at a Capitol Hill seminar on biotechnology sponsored by the Congressional Hunger Center, Shiva airily dismissed golden rice by claiming that "just in the state of Bengal 150 greens which are rich in vitamin A are eaten and grown by the women." A visibly angry Martina McGloughlin, director of the biotechnology program at the University of California at Davis, said "Dr. Shiva's response reminds me of . . . Marie Antoinette, [who] suggested the peasants eat cake if they didn't have access to bread." Alexander Avery of the Hudson Institute's Center for Global Food Issues noted that nutritionists at UNICEF doubted it was physically possible to get enough vitamin A from the greens Shiva was recommending. Furthermore, it seems unlikely that poor women living in shanties in the heart of Calcutta could grow greens to feed their children.

The apparent willingness of biotechnology's opponents to sacrifice people for their cause disturbs scientists who are trying to help the world's poor. At the annual meeting of the American Association for the Advancement of Science last February, Ismail Serageldin, the director of the Consultative Group on International Agricultural Research, posed a challenge: "I ask opponents of biotechnology, do you want 2 to 3 million children a year to go blind and 1 million to die of vitamin A deficiency, just because you object to the way golden rice was created?"

Vandana Shiva is not alone in her disdain for biotechnology's potential to help the poor. Mae-Wan Ho, a reader in biology at London's Open University who advises another activist group, the Third World Network, also opposes golden rice. And according to a *New York Times* report on a biotechnology meeting held last March by the Organization for Economic Cooperation and Development, Benedikt Haerlin, head of Greenpeace's European anti-biotech campaign, "dismissed the importance of saving African and Asian lives at the risk of spreading a new science that he considered untested."

Shiva, Ho, and Haerlin are leaders in a growing global war against crop biotechnology, sometimes called "green biotech" (to distinguish it from medical biotechnology, known as "red biotech"). Gangs of anti-biotech vandals with cute monikers such as Cropatistas and Seeds of Resistance have ripped up scores of research plots in Europe and the U.S. The so-called Earth Liberation Front burned down a crop biotech lab at Michigan State University on New Year's Eve in 1999, destroying years of work and causing $400,000 in property damage. (See "Crop Busters," January.) Anti-biotech lobbying groups have proliferated faster than bacteria in an agar-filled petri dish: In addition to Shiva's organization, the Third World Network, and Greenpeace, they include the Union of Concerned Scientists, the Institute for Agriculture and Trade Policy, the Institute of Science in Society, the Rural Advancement Foundation International, the Ralph Nader–founded Public Citizen, the Council for Responsible Genetics, the Institute for Food and Development Policy, and that venerable fount of biotech misinformation, Jeremy Rifkin's Foundation on Economic Trends. The left hasn't been this energized since the Vietnam War. But if the anti-biotech movement is successful, its victims will include the downtrodden people on whose behalf it claims to speak.

"We're in a war," said an activist at a protesters' gathering during the November 1999 World Trade Organization meeting in Seattle. "We're going to bury this first wave of biotech." He summed up the basic strategy pretty clearly: "The first battle is labeling. The second battle is banning it."

Later that week, during a standing-room-only "biosafety seminar" in the basement of a Seattle Methodist church, the ubiquitous Mae-Wan Ho declared, "This warfare against nature must end once and for all." Michael Fox, a vegetarian "bioethicist" from the Humane Society of the United States, sneered: "We are very clever simians, aren't we? Manipulating the bases of life and thinking we're little gods." He added, "The only acceptable application of genetic engineering is to develop a genetically engineered form of birth control for our own species." This creepy declaration garnered rapturous applause from the assembled activists.

Despite its unattractive side, the global campaign against green biotech has had notable successes in recent years. Several leading food companies, including Gerber and Frito-Lay, have been cowed into declaring that they will not use genetically improved crops to make their products. Since 1997, the European Union has all but outlawed the growing and importing of biotech crops and food. Last May some 60 countries signed the Biosafety Protocol, which mandates special labels for biotech foods and requires strict notification, documentation, and risk assessment procedures for biotech crops. Activists have launched a "Five-Year Freeze" campaign that calls for a worldwide moratorium on planting genetically enhanced crops.

For a while, it looked like the United States might resist the growing hysteria, but in December 1999 the Environmental Protection Agency announced that it was reviewing its approvals of biotech corn crops, implying that it might ban the crops in the future. Last May the Food and Drug Administration, which until now has evaluated biotech foods solely on their objective characteristics, not on the basis of how they were produced, said it would formulate special rules for reviewing and approving products with genetically modified ingredients. U.S. Rep. Dennis Kucinich (D-Ohio) has introduced a bill that would require warning labels on all biotech foods.

In October, news that a genetically modified corn variety called StarLink that was approved only for animal feed had been inadvertently used in two brands of taco shells prompted recalls, front-page headlines, and anxious recriminations. Lost in the furor was the fact that there was little reason to believe the corn was unsafe for human consumption—only an implausible, unsubstantiated fear that it might cause allergic reactions [more likely merely food intolerances—*Ed.*] Even Aventis, the company which produced Star-Link, agreed that it was a serious mistake to have accepted the EPA's approval for animal use only. Most proponents favor approving biotech crops only if they are determined to be safe for human consumption.

To decide whether the uproar over green biotech is justified, you need to know a bit about how it works. Biologists and crop breeders can now select a specific useful gene from one species and splice it into an unrelated species. Previously plant breeders were limited to introducing new genes through the time-consuming and inexact art of crossbreeding species that were fairly close relatives. For each cross, thousands of unwanted genes would be introduced into a crop species. Years of "backcrossing"—breeding each new generation of hybrids with the original commercial variety over several generations—were needed to eliminate these unwanted genes so that only the useful genes and characteristics remained. The new methods are far more precise and efficient. The plants they produce are variously described as "transgenic," "genetically modified," or "genetically engineered."

Plant breeders using biotechnology have accomplished a great deal in only a few years. For example, they have created a class of highly successful insect-resistant crops by incorporating toxin genes from the soil bacterium *Bacillus thuringiensis*. Farmers have sprayed B.t. spores on crops as an effective insecticide for decades. Now, thanks to some clever biotechnology, breeders have produced varieties of corn, cotton, and potatoes that make their own insecticide. B.t. is toxic largely to destructive caterpillars such as the European corn borer and the cotton bollworm; it is not harmful to birds, fish, mammals, or people.

Another popular class of biotech crops incorporates an herbicide resistance gene, a technology that has been especially useful in soybeans. Farmers can spray herbicide on their fields to kill weeds without harming the crop plants. The most widely used herbicide is Monsanto's Roundup (glyphosate), which toxicologists regard as an environmentally benign chemical that degrades rapidly, days after being applied. Farmers who use "Roundup Ready" crops don't have to plow for weed control, which means there is far less soil erosion.

Biotech is the most rapidly adopted new farming technology in history. The first generation of biotech crops was approved by the EPA, the FDA, and the

U.S. Department of Agriculture in 1995, and by 1999 transgenic varieties accounted for 33 percent of corn acreage, 50 percent of soybean acreage, and 55 percent of cotton acreage in the U.S. Worldwide, nearly 90 million acres of biotech crops were planted in 1999. With biotech corn, U.S. farmers have saved an estimated $200 million by avoiding extra cultivation and reducing insecticide spraying. U.S. cotton farmers have saved a similar amount and avoided spraying 2 million pounds of insecticides by switching to biotech varieties. Potato farmers, by one estimate, could avoid spraying nearly 3 million pounds of insecticides by adopting *B.t.* potatoes. Researchers estimate that *B.t.* corn has spared 33 million to 300 million bushels from voracious insects.

One scientific panel after another has concluded that biotech foods are safe to eat, and so has the FDA. Since 1995, tens of millions of Americans have been eating biotech crops. Today it is estimated that 60 percent of the foods on U.S. grocery shelves are produced using ingredients from transgenic crops. In April a National Research Council panel issued a report that emphasized it could not find "any evidence suggesting that foods on the market today are unsafe to eat as a result of genetic modification." *Transgenic Plants and World Agriculture*, a report issued in July that was prepared under the auspices of seven scientific academies in the U.S. and other countries, strongly endorsed crop biotechnology, especially for poor farmers in the developing world. "To date," the report concluded, "over 30 million hectares of transgenic crops have been grown and no human health problems associated specifically with the ingestion of transgenic crops or their products have been identified." Both reports concurred that genetic engineering poses no more risks to human health or to the natural environment than does conventional plant breeding.

As U.C.-Davis biologist Martina McGloughlin remarked at last June's Congressional Hunger Center seminar, the biotech foods "on our plates have been put through more thorough testing than conventional food ever has been subjected to." According to a report issued in April by the House Subcommittee on Basic Research, "No product of conventional plant breeding . . . could meet the data requirements imposed on biotechnology products by U.S. regulatory agencies. . . . Yet, these foods are widely and properly regarded as safe and beneficial by plant developers, regulators, and consumers." The report concluded that biotech crops are "at least as safe [as] and probably safer" than conventionally bred crops.

In opposition to these scientific conclusions, Mae-Wan Ho points to a study by Arpad Pusztai, a researcher at Scotland's Rowett Research Institute, that was published in the British medical journal *The Lancet* in October 1999. Pusztai found that rats fed one type of genetically modified potatoes (not a variety created for commercial use) developed immune system disorders and organ damage. *The Lancet*'s editors, who published the study even though two of six reviewers rejected it, apparently were anxious to avoid the charge that they were muzzling a prominent biotech critic. But *The Lancet* also published a thorough critique, which concluded that Pusztai's experiments "were incomplete, included too few animals per diet group, and lacked controls such as a standard rodent diet. . . . Therefore the results are difficult to interpret and do not allow the conclusion that the genetic modification of potatoes accounts for adverse effects in animals." The Rowett Institute, which does mainly nutritional research, fired Pusztai on the grounds that he had publicized his results before they had been peer reviewed.

Activists are fond of noting that the seed company Pioneer Hi-Bred produced a soybean variety that incorporated a gene—for a protein from Brazil nuts—that causes reactions in people who are allergic to nuts. The activists fail to mention that the soybean never got close to commercial release because Pioneer Hi-Bred checked it for allergenicity as part of its regular safety testing and immediately dropped the variety. The other side of the allergy coin is that biotech can remove allergens that naturally occur in foods such as nuts, potatoes, and tomatoes, making these foods safer.

Even if no hazards from genetically improved crops have been demonstrated, don't consumers have a right to know what they're eating? This seductive appeal to consumer rights has been a very effective public relations gambit for anti-biotech activists. If there's nothing wrong with biotech products, they ask, why don't seed companies, farmers, and food manufacturers agree to label them?

The activists are being more than a bit disingenuous here. Their scare tactics, including the use of ominous words such as *frankenfoods*, have created a climate in which many consumers would interpret labels on biotech products to mean that they were somehow more dangerous or less healthy than old-style foods. Biotech opponents hope labels would drive frightened consumers away from genetically modified foods and thus doom them. Then the activists could sit back and smugly declare that biotech products had failed the market test.

The biotech labeling campaign is a red herring anyway, because the U.S. Department of Agriculture plans to issue some 500 pages of regulations outlining what qualifies as "organic" foods by January, 2001. Among other things, the definition will require that organic foods not be produced using genetically modified crops. Thus consumers who want to avoid biotech products need only look for the "organic" label. Furthermore, there is no reason why conventional growers who believe they can sell more by avoiding genetically enhanced crops should not label their products accordingly, so long as they do not imply any health claims. The FDA has begun to solicit public comments on ways to label foods that are not genetically enhanced without implying that they are superior to biotech foods.

It is interesting to note that several crop varieties popular with organic growers were created through mutations deliberately induced by breeders using radiation or chemicals. This method of modifying plant genomes is obviously a far cruder and more imprecise way of creating new varieties. Radiation and chemical mutagenesis is like using a sledgehammer instead of the scalpel of biotechnology. Incidentally, the FDA doesn't review these crop varieties produced by radiation or chemicals for safety, yet no one has dropped dead from eating them.

Labeling nonbiotech foods as such will not satisfy the activists whose goal is to force farmers, grain companies, and food manufacturers to segregate biotech crops from conventional crops. Such segregation would require a great deal of duplication in infrastructure, including separate grain silos, rail cars, ships, and production lines at factories and mills. The StarLink corn problem is just a small taste of how costly and troublesome segregating conventional from biotech crops would be. Some analysts estimate that segregation would add 10 percent to 30 percent to the prices of food without any increase in safety. Activists are fervently hoping that mandatory crop segregation will also lead to novel legal nightmares: If a soybean shipment is inadvertently "contaminated" with biotech soybeans, who is liable? If biotech corn pollen falls on an organic cornfield, can the organic farmer sue the biotech farmer? Trial lawyers must be salivating over the possibilities.

The activists' "pro-consumer" arguments can be turned back on them. Why should the majority of consumers pay for expensive crop segregation that they don't want? It seems reasonable that if some consumers want to avoid biotech crops, they should pay a premium, including the costs of segregation.

As the labeling fight continues in the United States, anti-biotech groups have achieved major successes elsewhere. The Biosafety Protocol negotiated last February in Montreal requires that all shipments of biotech crops, including grains and fresh foods, carry a label saying they "may contain living modified organisms." This international labeling requirement is clearly intended to force the segregation of conventional and biotech crops. The protocol was hailed by Greenpeace's Benedikt Haerlin as "a historic step towards protecting the environment and consumers from the dangers of genetic engineering."

Activists are demanding that the labeling provisions of the Biosafety Protocol be enforced immediately, even though the agreement says they don't apply until two years after the protocol takes effect. Vandana Shiva claims the food aid sent to Orissa after the October 1999 cyclone violated the Biosafety Protocol because it was unlabeled. Greenpeace cited the unratified Biosafety Protocol as a justification for stopping imports of American agricultural products into Brazil and Britain. "The recent agreement on the Biosafety Protocol in Montreal . . . means that governments can now refuse to accept imports of GM crops on the basis of the 'precautionary principle,'" said a February 2000 press release announcing that Greenpeace activists had boarded an American grain carrier delivering soybeans to Britain.

Under the "precautionary principle," regulators do not need to show scientifically that a biotech crop is unsafe before banning it; they need only assert that it has not been proved harmless. Enshrining the precautionary principle into international law is a major victory for biotech opponents. "They want to err on the side of caution not only when the evidence is not conclusive but when no evidence exists that would indicate harm is possible," observes Frances Smith, executive director of Consumer Alert.

Model biosafety legislation proposed by Third World Network goes even further than the Biosafety Protocol, covering all biotech organisms and requiring authorization "for all activities and for all GMOs [genetically modified organisms] and derived products." Under the model legislation, "the absence of scientific evidence or certainty does not preclude the decision makers from denying approval of the introduction of the GMO or derived products." Worse, under the model regulations "any adverse socio-economic effects must also be considered." If this provision is adopted, it would give traditional producers a veto over innovative competitors, the moral equivalent of letting candlemakers prevent the introduction of electric lighting.

Concerns about competition are one reason European governments have been so quick to oppose crop biotechnology. "EU countries, with their heavily subsidized farming, view foreign agribusinesses as a competitive threat," Frances Smith writes. "With heavy subsidies and price supports, EU farmers see no need to improve productivity." In fact, a biotech-boosted European agricultural productivity would be a fiscal disaster for the E.U., since it would increase already astronomical subsidy payments to European farmers.

The global campaign against green biotech received a public relations windfall on May 20, 1999, when *Nature* published a study by Cornell University researcher John Losey that found that Monarch butterfly caterpillars died when force-fed milkweed dusted with pollen from *B.t.* corn. Since then, at every anti-biotech demonstration, the public has been treated to flocks of activist women dressed fetchingly as Monarch butterflies. But when more realistic field studies were conducted, researchers found that the alleged danger to Monarch caterpillars had been greatly exaggerated. Corn pollen is heavy and doesn't spread very far, and milkweed grows in many places aside from the margins of cornfields. In the wild, Monarch caterpillars apparently know better than to eat corn pollen on milkweed leaves.

Furthermore, *B.t.* crops mean that farmers don't have to indiscriminately spray their fields with insecticides, which kill beneficial as well as harmful insects. In fact, studies show that *B.t.* cornfields harbor higher numbers of beneficial insects such as lacewings and ladybugs than do conventional cornfields. James Cook, a biologist at Washington State University, points out that the population of Monarch butterflies has been increasing in recent years, precisely the time period in which *B.t.* corn has been widely planted. The fact is that pest-resistant crops are harmful mainly to target species—that is, exactly those insects that insist on eating them.

Never mind; we will see Monarchs on parade for a long time to come. Meanwhile, a spooked EPA has changed its rules governing the planting of *B.t.* corn, requiring farmers to plant non-*B.t.* corn near the borders of their fields so that *B.t.* pollen doesn't fall on any milkweed growing there. But even the EPA firmly rejects activist claims about the alleged harms caused by *B.t.* crops. "Prior to registration of the first *B.t.* plant pesticides in 1995," it said in response to a Greenpeace lawsuit, "EPA evaluated studies of potential effects on a wide variety of non-target organisms that might be exposed to the *B.t.* toxin, e.g., birds, fish, honeybees, ladybugs, lacewings, and earthworms. EPA concluded that these species were not harmed."

Another danger highlighted by anti-biotech activists is the possibility that transgenic crops will crossbreed with other plants. At the Congressional Hunger Center seminar, Mae-Wan Ho claimed that "GM-constructs are designed to invade genomes and to overcome natural species barriers." And that's not all. "Because of their highly mixed origins," she added, "GM-constructs tend to be unstable as well as invasive, and may be more likely to spread by horizontal gene transfer."

"Nonsense," says Tuskegee University biologist C. S. Prakash. "There is no scientific evidence at all for Ho's claims." Prakash points out that plant breeders specifically choose transgenic varieties that are highly stable since they want the genes that they've gone to the trouble and expense of introducing into a crop to stay there and do their work.

Ho also suggests that "GM genetic material" when eaten is far more likely to be taken up by human cells and bacteria than is "natural genetic material." Again, there is no scientific evidence for this claim. All genes from whatever source are made up of the same four DNA bases, and all undergo digestive degradation when eaten.

Biotech opponents also sketch scenarios in which transgenic crops foster superpests: weeds bolstered by transgenes for herbicide resistance or pesticide-proof bugs that proliferate in response to crops with enhanced chemical defenses. As McGloughlin notes, "The risk of gene flow is not specific to biotechnology. It applies equally well to herbicide resistant plants that have been developed through traditional breeding techniques." Even if an herbicide resistance gene did get into a weed species, most researchers agree that it would be unlikely to persist unless the weed were subjected to significant and continuing selection pressure—that is, sprayed regularly with a specific herbicide. And if a weed becomes resistant to one herbicide, it can be killed by another.

As for encouraging the evolution of pesticide-resistant insects, that already occurs with conventional spray pesticides. There is no scientific reason for singling out biotech plants. Cook, the Washington State University biologist, points out that crop scientists could handle growing pesticide resistance the same way they deal with resistance to infectious rusts in grains: Using conventional breeding techniques, they stack genes for resistance to a wide variety of evolving rusts. Similarly, he says, "it will be possible to deploy different *B.t.* genes or stack genes and thereby stay ahead of the ever-evolving pest populations."

The environmentalist case against biotech crops includes a lot of innuendo. "After GM sugar beet was harvested," Ho claimed at the Congressional Hunger Center seminar, "the GM genetic material persisted in the soil for at least two years and was taken up by soil bacteria." Recall that the *Bacillus thuringiensis* is a *soil bacterium*—its habitat is the soil. Organic farmers broadcast *B.t.* spores freely over their fields, hitting both target and nontarget species. If organic farms were tested, it's likely that *B.t.* residues would be found there as well; they apparently have not had any ill effects. Even the EPA has conceded, in its response to Greenpeace's lawsuit, that "there are no reports of any detrimental effects on the soil ecosystems from the use of *B.t.* crops."

Given their concerns about the spread of transgenes, you might think biotech opponents would welcome innovations designed to keep them confined. Yet they became apoplectic when Delta Pine Land Co. and the U.S. Department of Agriculture announced the development of the Technology Protection System, a complex of three genes that makes seeds sterile by interfering with the development of plant embryos. TPS also gives biotech developers a way to protect their intellectual property: Since farmers couldn't save seeds for replanting, they would have to buy new seeds each year.

Because high-yielding hybrid seeds don't breed true, corn growers in the U.S. and Western Europe have been buying seed annually for decades. Thus TPS seeds wouldn't represent a big change in the way many American and European farmers do business. If farmers didn't want the advantages offered in the enhanced crops protected by TPS, they would be free to buy seeds without TPS. Similarly, seed companies could offer crops with transgenic traits that would be expressed only in the presence of chemical activators that farmers could choose to buy if they thought they were worth the extra money. Ultimately, the market would decide whether these innovations were valuable.

If anti-biotech activists really are concerned about gene flow, they should welcome such technologies. The pollen from crop plants incorporating TPS would create sterile seeds in any weed that it happened to crossbreed with, so that genes for traits such as herbicide resistance or drought tolerance couldn't be passed on.

This point escapes some biotech opponents. "The possibility that [TPS] may spread to surrounding food crops or to the natural environment is a serious one," writes Vandana Shiva in her recent book *Stolen Harvest*. "The gradual spread of sterility in seeding plants would result in a global catastrophe that could eventually wipe out higher life forms, including humans, from the planet." This dire scenario is not just implausible but biologically impossible: *TPS is a gene technology that causes sterility; that means, by definition, that it can't spread.*

Despite the clear advantages that TPS offers in preventing the gene flow that activists claim to be worried about, the Rural Advancement Foundation International quickly demonized TPS by dubbing it "Terminator Technology." RAFI warned that "if the Terminator Technology is widely utilized, it will give the multinational seed and agrochemical industry an unprecedented and extremely dangerous capacity to control the world's food supply." In 1998 farmers in the southern Indian state of Karnataka, urged on by Shiva and company, ripped up experimental plots of biotech crops owned by Monsanto in the mistaken belief that they were TPS plants. The protests prompted the Indian government to declare that it would not allow TPS crops to enter the country. That same year, 20 African countries declared their opposition to TPS at a U.N. Food and Agriculture Organization meeting. In the face of these protests, Monsanto, which had acquired the technology when it bought Delta Pine Land Co., declared that it would not develop TPS.

Even so, researchers have developed another clever technique to prevent transgenes from getting into weeds through crossbreeding. Chloroplasts (the little factories in plant cells that use sunlight to produce energy) have their own small sets of genes. Researchers can introduce the desired genes into chloroplasts instead of into cell nuclei where the majority of a plant's genes reside. The trick is that the pollen in most crop plants don't have chloroplasts, therefore it is impossible for a transgene confined to chloroplasts to be transferred through crossbreeding.

As one tracks the war against green biotech, it becomes ever clearer that its leaders are not primarily concerned about safety. What they really hate is capitalism and globalization. "It is not inevitable that corporations will control our lives and rule the world," writes Shiva in *Stolen Harvest*. In *Genetic Engineering: Dream or Nightmare?* (1999), Ho warns, "Genetic engineering biotechnology is an unprecedented intimate alliance between bad science and big business which will spell the end of humanity as we know it, and the world at large." The first nefarious step, according to Ho, will occur when the "food giants of the North" gain "control of the food supply of the South through exclusive rights to genetically engineered seeds."

Accordingly, anti-biotech activists oppose genetic patents. Greenpeace is running a "No Patents on Life" campaign that appeals to inchoate notions about the sacredness of life. Knowing that no patents means no investment, biotech opponents declare that corporations should not be able to "own" genes, since they are created by nature.

The exact rules for patenting biotechnology are still being worked out by international negotiators and the U.S. Patent and Trademark Office. But without getting into the arcane details, the fact is that discoverers and inventors don't "own" genes. A patent is a license granted for a limited time to encourage inventors and discoverers to disclose publicly their methods and findings. In exchange for disclosure, they get the right to exploit their discoveries for 20 years, after which anyone may use the knowledge and techniques they have produced. Patents aim to encourage an open system of technical knowledge.

"Biopiracy" is another charge that activists level at biotech seed companies. After prospecting for useful genes in indigenous crop varieties from developing countries, says Shiva, companies want to sell seeds incorporating those genes back to poor farmers. Never mind that the useful genes are stuck in inferior crop varieties, which means that poor farmers have no way of optimizing their benefits. Seed companies liberate the useful genes and put them into high-yielding varieties that can boost poor farmers' productivity.

Amusingly, the same woman who inveighs against "biopiracy" proudly claimed at the Congressional Hunger Center seminar that 160 varieties of kidney beans are grown in India. Shiva is obviously unaware that farmers in India are themselves "biopirates." Kidney beans were domesticated by the Aztecs and Incas in the Americas and brought to the Old World via the Spanish explorers. In response to Shiva, C. S. Prakash pointed out that very few of the crops grown in India today are indigenous. "Wheat, peanuts, and apples and everything else—the chiles that the Indians are so proud of," he noted, "came from outside. I say, thank God for the biopirates." Prakash condemned Shiva's efforts to create "a xenophobic type of mentality within our culture" based on the fear that "everybody is stealing all of our genetic material."

If the activists are successful in their war against green biotech, it's the world's poor who will suffer most. The International Food Policy Research Institute estimates that global food production must increase by 40 percent in the next 20 years to meet the goal of a better and more varied diet for a world population of some 8 billion people. As biologist Richard Flavell concluded in a 1999 report to the IFPRI, "It would be unethical to condemn future generations to hunger by refusing to develop and apply a technology that can build on what our forefathers provided and can help produce adequate food for a world with almost 2 billion more people by 2020."

One way biotech crops can help poor farmers grow more food is by controlling parasitic weeds, an enormous problem in tropical countries. Cultivation cannot get rid of them, and farmers must abandon fields infested with them after a few growing seasons. Herbicide-resistant crops, which would make it possible to kill the weeds without damaging the cultivated plants, would be a great boon to such farmers.

By incorporating genes for proteins from viruses and bacteria, crops can be immunized against infectious diseases. The papaya mosaic virus had wiped out papaya farmers in Hawaii, but a new biotech variety of papaya incorporating a protein from the virus is immune to the disease. As a result, Hawaiian papaya orchards are producing again, and the virus-resistant variety is being made available to developing countries. Similarly, scientists at the Donald Danforth Plant Science Center in St. Louis are at work on a cassava variety that is immune to cassava mosaic virus, which killed half of Africa's cassava crop two years ago.

Another recent advance with enormous potential is the development of biotech crops that can thrive in acidic soils, a large proportion of which are located in the tropics. Aluminum toxicity in acidic soils reduces crop productivity by as much as 80 percent. Progress is even being made toward the Holy Grail of plant breeding, transferring the ability to fix nitrogen from legumes to grains. That achievement would greatly reduce the need for fertilizer. Biotech crops with genes for drought and salinity tolerance are also being developed. Further down the road, biologist Martina McGloughlin predicts, "we will be able to enhance other characteristics, such as growing seasons, stress tolerance, yields, geographic distribution, disease resistance, [and] shelf life."

Biotech crops can provide medicine as well as food. Biologists at the Boyce Thompson Institute for Plant Research at Cornell University recently reported success in preliminary tests with biotech potatoes that would immunize people against diseases. One protects against Norwalk virus, which causes diarrhea, and another might protect against the hepatitis B virus which afflicts 2 billion people. Plant-based vaccines would be especially useful for poor countries, which could manufacture and distribute medicines simply by having local farmers grow them.

Shiva and Ho rightly point to the inequities found in developing countries. They make the valid point that there is enough food today to provide an adequate diet for everyone if it were more equally distributed. They advocate land reform and microcredit to help poor farmers, improved infrastructure so farmers can get their crops to market, and an end to agricultural subsidies in rich countries that undercut the prices that poor farmers can demand.

Addressing these issues is important, but they are not arguments against green biotech. McGloughlin agrees that "the real issue is inequity in food distribution. Politics, culture, regional conflicts all contribute to the problem. Biotechnology isn't going to be a panacea for all the world's ills, but it can go a long way toward addressing the issues of inadequate nutrition and crop losses." Kenyan biologist Florence Wambugu argues that crop biotechnology has great potential to increase agricultural productivity in Africa without demanding big changes in local practices: A drought-tolerant seed will benefit farmers whether they live in Kansas or Kenya.

Yet opponents of crop biotechnology can't stand the fact that it will help developed countries first. New technologies, whether reaping machines in the 19th century or computers today, are always adopted by the rich before they become available to the poor. The fastest way to get a new technology to poor people is to speed up the product cycle so the technology can spread quickly. Slowing it down only means the poor will have to wait longer. If biotech crops catch on in the developed countries, the techniques to make them will become available throughout the world, and more researchers and companies will offer crops that appeal to farmers in developing countries.

Activists like Shiva subscribe to the candlemaker fallacy: If people begin to use electric lights, the candlemakers will go out of business, and they and their families will starve. This is a supremely condescending view of poor people. In order not to exacerbate inequality, Shiva and her allies want to stop technological progress. They romanticize the backbreaking lives that hundreds of millions of people are forced to live as they eke out a meager living off the land.

Per Pinstrup-Andersen of the International Food Policy Research Institute asked participants in the Congressional Hunger Center seminar to think about biotechnology from the perspective of people in developing countries: "We need to talk about the low-income farmer in West Africa who, on half an acre, maybe an acre of land, is trying to feed her five children in the face of recurrent droughts, recurrent insect attacks, recurrent plant diseases. For her, losing a crop may mean losing a child. Now, how can we sit here debating whether she should have access to a drought-tolerant crop variety? None of us at this table or in this room [has] the ethical right to force a particular technology upon anybody, but neither do we have the ethical right to block access to it. The poor farmer in West Africa doesn't have any time for philosophical arguments as to whether it should be organic farming or fertilizers or GM food. She is trying to feed her children. Let's help her by giving her access to all of the options. Let's make the choices available to the people who have to take the consequences."

# Organic or Genetically Modified Food Which Is Better?

## 11

GREGORY E. PENCE

*Gregory E. Pence has been a member of the University of Alabama–Birmingham (UAB) faculty since January 1976. In addition to teaching in the Department of Philosophy, he teaches a course in medical ethics to all first-year students in UAB's School of Medicine. Pence graduated cum laude in philosophy in 1970 from the College of William and Mary in Williamsburg, Virginia, and earned his Ph.D. in 1974 from New York University.*

*Pence's books include* Re-Creating Medicine: Ethical Issues at the Frontiers of Medicine, Classic Cases in Medical Ethics: Accounts of the Cases That Shaped Medical Ethics, Who's Afraid of Human Cloning?, *and, with G. Lynn Stevens,* Seven Dilemmas in World Religions. *He has edited* Classic Works in Medical Ethics: Core Philosophical Readings *and* Flesh of My Flesh: The Ethics of Cloning Humans.

*This essay, from Pence's 2002 book* Designer Food: Mutant Harvest or Breadbasket of the World?, *argues that GM foods are actually safer than organic foods.*

> *You don't have to sacrifice style or comfort for an organic lifestyle. It's all about making choices that make a difference in the world—making it cleaner, greener, and healthier.*
>
> —MARIA RODALE, editor, *Organic Style* magazine, and author, *The Organic Suburbanite*

IN THE MODERN WORLD, no foods could be more evaluatively distinct than genetically modified (GM) food and organic food. GM food is unnatural, dangerous, and grown by international conglomerates; organic food is natural, safe, wholesome, and grown locally on small farms. GM food departs

---

From Gregory E. Pence, *Designer Food: Mutant Harvest or Breadbasket of the World?* (Lanham, Md.: Rowman & Littlefield, 2002).

dangerously from traditional practices, whereas organic food improves on tainted, industrial norms by returning to purer, simpler ways of cultivating food. At least, these are our myths.

In truth, organic food can be unsafe. Organic lettuce or spinach, commonly grown in soil to which manure has been added, can contain *E. coli* bacteria, which can cause hemorrhagic colitis, acute kidney failure, and even death. The Centers for Disease Control estimated that food-borne illnesses caused by the O157:H7 strain of *E. coli* sickened over 73,480 Americans in 1999.[1] Although the etiology of this strain of *E. coli* is poorly understood, in 1993 it was known to have infected more than seven hundred people in western states, killing four, because of undercooked hamburgers served by the Jack-in-the-Box fast-food chain.[2] Children are especially vulnerable to it.

Ironically, eating organic food may expose people to more, not fewer, risks from O157:H7. Outbreaks of *E. coli* O157:H7 stem not only from bad beef, but also from fresh fruit juices, raw milk, lettuce, and minimally processed produce. Unpasteurized apple juice in the Pacific Northwest killed a child and sent sixty-six others to the hospital in the fall of 1996.[3] In 2000, the Tesco supermarket chain pulled from its shelves all organic mushrooms because they tested positive for *E. coli* O157:H7. Putting the lie to the myth that all organic food is local, Tesco said that the suspect batch came from a supplier in Northern Ireland that used Belgian manure.[4]

Understandably, the Beef Industry Food Safety Council wants the public to understand that beef is not the major culprit here. Indirectly referring to the outbreak at the Jack-in-the-Box franchises, the council emphasizes:

> While outbreaks attributed to beef grab headlines, perhaps more attention should be given to fruits and vegetables. The CDC report on surveillance of foodborne disease outbreaks, also released in March 2000, showed that the number of cases of foodborne-illnesses in outbreaks attributed to fruits and vegetables have exceeded those of beef every single year, even by 10-to-1, as in 1995.[5]

So how does organic food become infected with *E. coli*? First, organic food is usually grown in soil with manure, a natural breeding ground for various forms of *E. coli*, as well as any other chemical that might be passed through an animal's body. Although it is a bit gross to dig too deeply into the nature of this pile, one wonders how the assumption got accepted that food grown in manured soil is pure while food that is grown with genes that help plants fix nitrogen from the soil is bad.

Second, growing organic food is labor intensive (Prince Charles's organic garden requires a dozen full-time gardeners) and commercial growers employ unskilled workers to reduce costs. Yuppies who eat organic food now do so because of hundreds of thousands of low-paid workers toiling in back-breaking jobs. Nothing about hand weeding or retilling rows to avoid weeds is easy. Natural ways of growing food are usually primitive and require enormous amounts of human capital.

Ridding organic produce of manure also requires careful washing and such work is boring. But the safety of this food rests on the care of this work. In one case of infected organic produce, bacteria contaminated the water used to rid the produce of manure.[6]

Prima facie, the concepts of organic food and genetically modified food would seem to occupy different universes. Don't these different kinds of food involve groups of plants that share nothing in common? People who believe this, in truth, oversimplify.

Plants that exist in nature generate their own chemicals for killing unwanted insects. Consider rotenone, which comes from the roots and leaves of tropical plants such as jewel vine, derris, and the hoary pea. Commonly sold as a white powder, this potent stuff kills fire ants and is used by organic farmers on tomatoes, pears, and apples. Horticulturists spray it on roses and African violets. Because this pesticide occurs inside a plant that evolved naturally in evolution, it is considered permissible to use it in nearly seven hundred products used in organic gardens.

Scientists at Emory University in Atlanta discovered in 2000 that rotenone injected into rats caused Parkinson's disease–like symptoms.[7] They already knew that some environmental toxin was at work here because farmworkers were seven times more likely than normal to develop Parkinson's.

I am not arguing here that the dangers of organic foods surpass those of GM foods. Rather, I want to expose the naïve assumption that a pesticide (rotenone) created internally by exotic, tropical plants and then manufactured in great quantities and sprayed on traditional food crops is somehow more safe, more benevolent, and less artificial than plants to which a few benevolent genes have been added to the existing thirty thousand.

I do not believe that everything called "natural" is safe and good. Cassava, a natural crop grown in most countries in Africa, when improperly prepared causes acute cassava poisoning, TAN (tropical ataxic neuropathy), Konzo (upper motor neuron paraparesis), and goiter.[8]

Besides, and in a more general sense, who decides what is "natural"? In almost every way, what we commonly call natural is a social construction—in other words, a judgment at a particular time and by a particular culture. In the national parks in North America, the asphalt roads, firefighters who fight lightning fires, forest rangers, pest management, and suppression of dangers to humans like mountain lions, bears, and wolves don't really reflect nature. Many of the plants we take as natural to North America, such as Japanese honeysuckle and purple loosestrife, came as imports in the eighteenth or nineteenth centuries.

Critics admit that millions of Americans have already eaten products from genetically modified corn and soybeans, and while they foresee botanical Armageddon from such modifications, no North American has even been sickened (much less killed!) by eating such GM veggies. We take our system of producing meat as normal and safe, yet it may be far more dangerous than any GM foods.

It is neither splitting hairs nor asking rhetorical questions to argue that GM veggies and organic crops resemble each other more than they differ. Take the infamous Bt corn, alleged to endanger the Monarch butterfly. *Bacillus thuringiensis*, a.k.a. "Bt," occurs naturally in soils around the world. Consumers Union calls this bacterium, a natural pesticide, "more benign than many synthetic [chemical] pesticides."[9] It produces a crystalline protein (a "cry" protein) that kills pests on farms, especially the European corn borer. (Note that this corn borer is a close relative of a caterpillar that becomes the Monarch butterfly, so the U.S. Department of Agriculture and the Food and Drug Administration (FDA) knew that Bt pollen could affect such larvae, especially when artificially dumped on larvae in artificial conditions, but then again, so could Bt as a concentrated spray.)

Because Bt occurs everywhere in soil, and because it does not damage good insects (honeybees, ladybugs) that help control bad ones, it can be sprayed on crops and still have the crops labeled "organic." However, this Bt pesticide must be sprayed and resprayed on crops to achieve the desired effects.

Enter GM corn. Bt corn is nothing more than traditional corn containing inside it the genes of *Bacillus thuringiensis*. In other words, if a plant has been soaked with Bt spray over and over for months, and hence absorbs Bt into its cells, it can still be called "organic," but if we insert Bt genes into the corn and no spray is used, then the corn becomes dangerous, as it's been "genetically modified."

Obviously, organic food in this case does not differ as much from genetically modified food as associations of organic farmers would urge us to believe. No longer conducted only on small farms, organic farming now mimics industrial farming, with some plants imported from huge overseas farms and processed together in huge factories. In the making of what the organic industry calls "an organic TV dinner" (if that is not an oxymoron, what is?), "broccoli is trucked to Alberta, Canada, . . . to meet up with pieces of organic chicken that have already made a stop at a processing plant in Salem, Oregon, where they were defrosted, injected with marinade, cooked, and refrozen."[10]

Because consumers turn to organic foods when they become scared about conventional food, organic growers profit from scares about genetically modified food; e.g., sales of their produce in England soared five-fold during the five years of scares about mad cow disease.[11] Some defenders of traditional agriculture see a well-orchestrated campaign by the natural food industry to scare consumers about GM foods.

At least we know that organic food is better for the world's environment, right? Not so. Indeed, just the opposite. Chemical fertilizers that utilize nitrogen in the atmosphere cannot be used in organic farming, which instead must use animal manure for crops that don't fix their own nitrogen. To fertilize organic crops, animal and human waste, mainly the manure of cattle, must be used. A major supplier to England and Europe, Icelandic Incorporated, destroys forests and grasslands in Ecuador to grow more organic food.

Nobel prize–winning plant biologist Norman Borlaug laughs at the proposal that the world can be fed on organic foods:

> At the present time, approximately 80 million tons of nitrogen nutrients are utilized each year. If you tried to produce this nitrogen organically, you would require an additional 5 or 6 billion head of cattle to supply the manure. How much wild land would you have to sacrifice just to produce the forage for these cows? There's a lot of nonsense going on here. . . .
>
> As far as plants are concerned, they can't tell whether that nitrate ion comes from artificial chemicals or from decomposed organic matter. If some consumers believe that it's better from the point of view of their health to have organic food, God bless them. Let them buy it. Let them pay a bit more. It's a free society. But don't tell the world that we can feed the present population without chemical fertilizer. That's when this misinformation becomes destructive.[12]

But surely eating organic is more nutritional, right? To quote Borlaug again, "If people want to believe that the organic food has better nutritive value, it's up to them to make that foolish decision. But there's absolutely no research that shows that organic foods provide better nutrition."

Simple-minded assumptions about nature, food, health, and the environment underlie people's embrace of organic foods. Chef Jim White, in his "Kitchen Journal" column in the *Albuquerque Journal*, says, "As I've always said, if you want to be healthy, you have to stay as close to nature as possible. When an animal or plant is raised in an organic environment, you are giving your body what Mother Nature intended it to have."[13]

Chef Jim capitalizes "Mother Nature" to indicate his romantic naturalism. An invisible, Divine Hand guided agriculture of the early nineteenth century, such that the farms of Ohio in 1840 were perfect. If only we could get back to that hardscrabble life, he seems to think, we would all live to be a hundred.

Britain's royal family also debates organic and GM foods. Prince Charles, a critic of GM food, champions organic farming, claiming that it aligns with a vague deity whom Charles variously calls the "Sustainer" or "Creator."[14] Such a deity, Charles assures us, frowns on genetically modified food crops. To grow GM food "usurps our place in Nature." Presumably, growing organic crops does not, but that also implies that tedious hand weeding and spreading of cow manure is our natural place.

Prince Phillip and Princess Anne support genetically modified crops, accusing Charles of "oversimplifications." As Prince Phillip corrected his son publicly, "Do not forget that we have been genetically modifying animals and plants ever since people started selective breeding. People are worried about genetically modified organisms getting into the environment. What people forget is that the introduction of exotic species—like, for instance, the gray squirrel into this country—is going to, or has done, far more damage than a genetically modified piece of potato."[15]

Most scientists agree here with Prince Phillip and Princess Anne. Thousands of plants have been created by being crossbred in traditional ways in which genes randomly mix. Such plants (seedless grapes, tangelos) enter our environment all the time with little testing or regulation. Newly imported foods are also introduced in our grocery stores and restaurants (Kiwi fruit, sushi).

Unless some reason arises to think that a newly introduced food has a protein to which humans are allergic, it is not subjected to the same exhaustive battery of tests that GM veggies undergo. GM veggies get tested more than some of the new food we eat in restaurants or some of the exotic fruits and vegetables that suddenly appear in our grocery stores.

Overall, our present system for testing food is more than a bit hypocritical about the safety of traditional food and more than a bit hysterical about genetically modified veggies. As always, people fear the new kid on the block, especially if his name is "Gene."

# Notes

1. www.cdc.gov/ncidod/eid/vol1no2/feng.htm. For documentation from the beef industry, especially about the greater dangers of *E. coli* O157:H7 from fruits and vegetables, see www.beef.org/library/factsheets/fs_e_coli.htm.

2. Peter Feng, "Escherichia coli Serotype O157:H7: Novel Vehicles of Infection and Emergence of Phenotypic Variants," Centers for Disease Control, www.cdc.gov/ncidod/eid/vol1no2/feng.htm.

3. Barry Yeoman, "Dangerous Food," *Redbook*, August 2000, 123.

4. www.thisislondon.co.uk/dynamic/news/business/top_direct.

5. Christina Pope, "Fact Sheet: E. coli O157:H7," Beef Industry Food Safety Council (BIFSCo), www.beef.org/library/factsheets/fs_e_coli.htm.

6. "Warning: Organic and Natural Foods May Be Hazardous to Your Health," Knight Ridder/Tribune News Service, 2 May 2000 (Internet edition).

7. Sandra Blakeslee, "Pesticide Found to Produce Parkinson's Symptoms in Rats," *New York Times*, 5 November 2000 (Internet edition).

8. From a posting by medical toxicologist Alan H. Hall on the AgBio Listserv, dated May 9, 2000.

9. "Seeds of Change," *Consumers Reports*, September 1999, 43.

10. Michael Pollan, "How Organic Became a Marketing Niche and Multi-billion-Dollar Industry," *New York Times Magazine*, 13 May 2001, 32.

11. Marian Burros, "Mainstream Organics: Britain Stocks Up," *New York Times*, 26 June 2000, B11 (national ed.).

12. Norman Borlaug, "Taking the Gm Food Debate to Africa: Have We Gone Mad?," open letter to the editor, *Independent Newspaper*, London, 10 April 2000.

13. Jim White, "Organic Foods More Healthful," *Albuquerque Journal*, 2 August 2000, C1.

14. James Meek, "Prince Courts Controversy as He Places the Nature of God above the God of Science," *Guardian*, 17 May 2000, Internet edition.

15. James Meek, "Duke Challenges Sceptics over GM Food," *Guardian*, 7 June 2000 (Internet edition).

# The Benefits of Organic Food          12

TANYA MAXTED-FROST

*Tanya Maxted-Frost is a journalist and writer based in London who writes about natural foods and organic cooking. She wrote* The Organic Baby Book *and coauthored with Daphne Lambert* The Organic Baby and Toddler Cookbook *(both published by Green Books).*

*In the essay that follows, Maxted-Frost emphasizes the many residues from pesticides and waxes that cover nonorganic produce. She champions eating produce and meat free of ingredients from GM plants and distrusts politicians who claim that GM food is safe.*

THE FRONT PAGE NEWS STORY in June that human breastmilk has been found to contain over 350 manmade chemicals which are being passed on to breastfed babies in increased toxic concentrations[1] may have shocked many people. It may unfortunately have caused some mothers to stop breastfeeding their infants, and others to avoid it altogether, despite official pleas that breast is still best. (It's a shame there was no similar report available at the same time into pollutants in formula milks, dairy cows and their milk.) But while the WWF-UK report behind the headlines may have been new, the tragedy it told wasn't.

Many warnings of this toxic chemical legacy being passed from generation to generation have been voiced over the past 50 years by scientists, organic and environmental pioneers and activists who foresaw the dangers or witnessed first hand the evidence of it in both animals and humans around the world. Worryingly many of these experts and scientists, such as Liverpool University foetal toxicologist Dr. Vyvyan Howard, say and have the evidence to show that the intrauterine effect of these chemicals (even in small amounts and especially in combination) is even more toxic and harmful to the develop-

---

This essay has been reprinted with permission from Issue 47 *Positive Health* magazine (Dec. 1999)—www.positivehealth.com.

ing embryo and foetus. The damage is already being done before our babies are born.

Published warnings made available to the public have included Rachel Carson's seminal book *Silent Spring* in 1962, Dr. Theo Colborn's *Our Stolen Future* in 1996 and Professor John Wargo's *Our Children's Toxic Legacy* in 1998. Their collective pleas for official sanity have largely fallen on deaf ears and so the legacy—which stretches to the furthest reaches of the earth affecting Eskimos, polar bears (some of which have been found to be hermaphrodites), seals, dolphins and whales alike[2]—has continued unabated.

So what does all of this have to do with the food we eat? Surely the sources of most of these pollutants contaminating our breastmilk and body fat are beyond our control—the fallout from years of industrial waste pollution from factories and production of PVC plastics and other consumer goods, and from traffic pollution, etc.? Think again. The food we eat—something we can control—is a major contributor. Most of our food is produced on an industrial scale, using artificial fertilisers—up to two-thirds of which leach into and contaminate groundwater, lakes and streams[3]—and pesticide, herbicide and fungicide chemicals which carry poison/hazardous warnings on their containers and require protective clothing and a licence when used. In 1997 alone, over 25 million kilograms of these chemicals were sold in the UK and most were sprayed onto Britain's fields.[4]

Most of these are not biodegradable and persist to contaminate the environment, kill wildlife (many birds, including three quarters of the skylark population, have vanished from the countryside due to intensive farming practices[5]), contaminate our drinking water (millions of pounds is spent each year trying to remove pesticides by water companies[6]) and leave detectable residues in or on our foods and drinks.

Our daily diet is now reported to contain residues of some 30 different artificial chemicals[7] and a supermarket apple may have been treated up to 40 times with any of 100 chemicals.[8] The government warns us to peel fruits such as apples and pears and peel, top and tail carrots due to ongoing detection of these residues, and the wide variance in residue levels. It's now thought that one in 1000 apples could give you a headache or stomach upset.[9] UK farmers are even occasionally found to use chemicals banned for use in this country on produce grown for human consumption, and imported foods can also contain residues of chemicals from the UK banned list.[10] Many of these pesticides are known carcinogens and are also linked to other degenerative diseases such as Alzheimer's and Parkinson's. Some are also hormone disrupters, mimickers or blockers—able to trigger hormone-related diseases.[11]

The feed of livestock and poultry bound for our dinner tables as meat can contain pesticides, routine antibiotics (contributing to the current looming crisis of resistant super bugs in both animals and humans, and the increasing loss of effectiveness of lifesaving antibiotics), colourants (to help battery hens produce yellow yolks), growth hormones, and even engine oil (remember the recent Belgian foods debacle?) and stray dead dogs (a practice stumbled upon in Ireland in livestock bound for the EU).[12]

Farmed fish are also fed colourants (to turn the flesh pink) and drugs to prevent outbreaks of disease, which invariably reach wild waterways.[13] And will we ever forget BSE and CJD? And what about genetically engineered foods—GM ingredients are already found in 60 per cent of non-organic processed foods and most will remain unlabelled despite new EU legislation.[14]

A recent Food Commission survey also revealed that there can be up to two per cent GM ingredients in non-organic products labelled 'GM free' and that the EU is likely to set a two to three per cent tolerance level. New toxins are feared from these new, untested (on human health) foods which have been found to be toxic to many insects and have foreign genes transplanted into them from other plant or animal species and viruses which are used to 'switch' the new genes on.[15] GM food production also poses the biggest threat to organic food production through contamination by cross-pollination—many campaigners fear that the two cannot co-exist on an island the size of Britain.[16]

And while we make all this 'progress' with hi-tech food production and farming, cancers, other degenerative diseases and infertility are rising and sperm counts and nutrient levels in our farming soils and food are declining.[17] The NHS groans with the burden of more and more sick and malnourished people, and the lists grow long of those awaiting operations.

Key mineral and vitamin deficiencies are becoming widespread[18]— thought to play a vital part in the increase in poor health, behavioural problems and even in the increasing number of criminals. Preventative health is not a popular term in this country. Through our taxes we're paying millions of pounds each year to support unsustainable farming and food, to clean up the mess of fiascos such as BSE, to detect residues of harmful substances which shouldn't be put on or in our food in the first instance, to have our drinking water cleaned of farming pollutants (many of which still remain), to set up a food agency and yet more committees to monitor and regulate conventional food processing (when it's the production system which is the real problem) and worrying developments such as GM foods, and to support an ailing health service treating an ever increasing number of polluted people at the end of the polluted food chain.

Surely we have the basic right to be able to buy and eat uncontaminated food free of artificial chemical additives and GM ingredients? Surely our babies and our children have the right not to be force fed these substances or poisoned by them in the womb, on the breast or in their weaning solids? So where do we turn? To the politicians and government agencies who say there are no risks and it's all perfectly safe (as they did during the BSE crisis)? Just what is safe for us to eat and to feed our families?

Organic and biodynamic foods, including those grown under permaculture systems, hold the answer. Their benefits include the fact that they are grown as naturally as possible without the aforementioned unnecessary additives, and that they are processed without others such as hydrogenated fats, modified starches, artificial flavourings, colourings, sweeteners, preservatives and most other E numbers commonly found in many non-organic foods. Instead of being intensively farmed unsustainably so that natural resources are depleted, and the traditional hedgerow landscape destroyed and turned into monoculture prairies, organic lands and soils are nurtured and built up naturally and gradually with plant and animal manures, composts and rotation methods to increase their fertility—the way they traditionally were before the 'grow big and quick at all costs' chemical methods were introduced.

Scientists and organic pioneers such as the great late Sir Albert Howard promoted the human health benefits of foods farmed this way as far back as the late 1930s and early 1940s when he warned about the dangers of going down the chemical additive route with farming and the public's food, telling us (based on decades of his international work and studies) that: 'A fertile soil means healthy crops, healthy livestock and last but not least healthy human beings. Soil fertility is the basis of the public health system of the future.'[19] Lady Eve Balfour, founder of the Soil Association in the late 1940s, similarly said 'The soil is a living organism. The health of man, beast, plant and soil is one indivisible whole.'[20] Nutrition pioneers of the 1930s, Drs Weston Price and Francis Pottenger of the US, and Sir Robert McCarrison of the UK, showed through their extensive studies of humans and animals on wholesome diets natural to them versus refined diets unnatural to them that good health depends on good nutrition.[21]

How sad then that all this early foresight on the need for good, healthy, unadulterated food was officially ignored and that instead organic organisations such as the Soil Association have had to struggle with limited funds and continually battle authorities to have organic food and farming recognised for its merits and benefits to human health and the environment. Organic farming still makes up less than one per cent of total UK agricultural land;[22] and people have to march, protest and forcefully demand what should be our birthright—the right to pure, wholesome, affordable, unadulterated healthy food.[23] Also, an organisation as valuable to healthy future generations as the charity the Foresight Association for the Promotion of Pre-conceptual Care remains relatively unknown in the public eye and poorly funded. It helps to get thousands of couples around the world each year onto healthy organic wholefood diets as part of its programme four to six months before they conceive in order to produce a healthy baby.

The successful Gerson therapy for battling cancer only uses organic foods and claims the therapy simply doesn't work with non-organic produce.[24] Men eating organic diets have been shown to have higher sperm counts than those who don't,[25] and Foresight has research results showing a ninety percent success rate, with most of its couples following its organic food programme going on to have healthy, ideal weight babies—even for those couples who previously were 'infertile'.[26] By comparison IVF's average success rate is 16.7%.[27]

There are countless positive anecdotes from individuals and families in books such as *The Shopper's Guide to Organic Food* by Lynda Brown and *The Organic Baby Book* who have converted to organic foods and noticed a marked improvement in their own and their families' health.

Anyone who eats a mainly organic diet wouldn't dream of going back intentionally. While we await the scientific evidence to prove what we already know about organic wholefoods and their benefits, and as the supply and quality continue to increase (there is now three times the amount of UK land in conversion than there is existing as organic[28]) the jury is in session daily on the case of non-organic food and farming and its detrimental effects on our health and our environment, collecting more and more damning evidence against it.

It's time the judge finally ordered its demise.

# References

1. Chemical Trespass: A Toxic Legacy, a WWF-UK Toxics Programme Report reviewing various international scientific studies and World Health Organization findings, including

Jensen AA and Slorach SA. Chemical contaminants in human milk. CRC Press, Florida. 1991; MAFF. Dioxins and Polychlorinated biphenyls in foods and human milk. Food surveillance information sheet, Number 105, Jun 1997.

2. Colborn Theo. Our Stolen Future. Abacus. 1997. Beland P, De Guise S, Girard C, Lagacee A, Martinequ D, Michaud R, Muir D, Norstrom R, Pelletier E, Ray S and Shugart L. Toxic Compounds and Health and Reproductive Effects in St. Lawrence Beluga Whales. Journal of Great Lakes Research 19(4): 766–75. 1993.

3. The Greening of the Green Revolution. Nature 19 Nov 1988.

4. Buffin David. Pesticides Trust: The SAFE Alliance, Food Indicator's Report. Datamonitor, Natural and Organic Food and Drinks. 1999.

5. Official communication. Royal Society for the Protection of Birds. 21 Mar 1999.

6. Pretty Jules. The Living Land. 1998.

7. Working Party on Pesticide Residues Annual Report, 1998. Brown Lynda. The Shopper's Guide to Organic Food. Fourth Estate 1998.

8. The True Cost of Food. Greenpeace and Soil Association, 1999; The Times, Daily Mail 1999; Dr. Sarah Brewer. Quoted in Maxted-Frost Tanyia. The Organic Baby Book, Green Books 1999.

9. Pesticides News. Pesticides Trust. Number 39.

10. Working Party on Pesticide Residues, Annual Report 1999.

11. Colborn T et al. Developmental effects of endocrine disrupting chemicals in wildlife and humans. Environ Health Prospect. 101:378–84. 1993. Various Pesticides Trust Publications 1998 and 1999.

12. Various including Soil Association reports; various speakers at the Compassion in World Farming conference, London 1998; Penman Danny. The Price of Meat. 1996; Harvey Graham, speaker, The Organic Food & Wine Festival, London, Aug 1, 1999.

13. Personal communication, conventional fish farmer turned organic, 1998.

14. Personal communication, The Soil Association September 1999; Greenpeace official communications 1999; Food Magazine July/September 1999.

15. Personal communication, GE expert Dr Michael Antoniou of Guys Hospital, London and Dr Vyvyan Howard, Liverpool University foetal toxicologist September 1999; Genetic Engineering Network internationally sourced scientific studies, reports, etc. email service, 1999.

16. Keenan Lindsay. Greenpeace, speaking at The Organic Food & Wine Festival, London, August 1, 1999.

17. Carlsen E, Giwercman A, Keiding N and Skakkeback NE. Evidence for Decreasing Quality of Semen During Past 50 Years. BMJ. 305: 609–13. 1992. Various including personal communication with Dr Marilyn Glenville, Leslie Kenton and Patrick Holford, 1999.

18. Personal communication, Leslie Kenton, 1999; Antony Haynes, The Nutrition Clinic speaking at The Organic Food & Wine Festival, August 1, 1999. The Case for Use of Micronutrient Food Supplements in the UK—Evidence of Reduced Population Intakes. Lamberts Nutrition Bites. Issue 5. 1998.

19. Soil and Health. Sir Albert Howard Memorial Number. Spring, 1948.

20. Balfour Eve. The Living Soil. 1946.

21. Price Weston A. Nutrition and Physical Degeneration. Price-Pottenger Nutrition

Foundation. La Mesa, CA. 1945. Pottenger F. M. Jr. Pottenger's Cats. Price-Pottenger Nutrition Foundation. La Mesa, CA. 1983. McCarrison Sir Robert. Nutrition and Health. McCarrison Society. London. 1984.

22. The Organic Food and Farming Report. The Soil Association. 1999.

23. Smith Bob L. Organic Foods vs Supermarket Foods: Element Levels. Journal of Applied Nutrition. 45(1). 1993. Campden Food and Drink Research Association study. 1998.

24. Personal communication with Gerson approved therapists, 1999.

25. Abell A, Ernst E and Bonde JP, High sperm density among members of organic farmers association. Lancet. 343: 1498. 1994.

26. Bradley SG and Bennett N. Preparation for Pregnancy, an Essential Guide. Argyll. 1997.

27. Human Fertilisation and Embryology Authority Figures, 1998.

28. The Organic Food and Farming Report 1999.

# Genetic Engineering and Food Security 13

VANDANA SHIVA

*A native of India, Vandana Shiva is a writer and science policy advocate known as a critic of biotechnology and genetically modified food and as an ecofeminist. She is director of the Research Foundation for Science, Technology and Natural Resource Policy, which criticizes dominant Western views of agriculture, commerce, and science, and vice president of the Third World Network. Her current work centers on biodiversity and sustainable agriculture.*

*Shiva is the author of* Biopiracy: The Plunder of Nature and Knowledge, Staying Alive, The Violence of the Green Revolution, *and* Monocultures of the Mind *and an associate editor of* The Ecologist. *Shiva won the Right Livelihood Award in 1993, which egalitarians call the "Alternative Nobel Peace Prize."*

*In this selection, Shiva argues that genetically modified food and patents on genes benefit huge, Western corporations, not ordinary people.*

GENETIC ENGINEERING has been sold as a green technology that will protect nature and biodiversity. However, the tools of genetic engineering are designed to steal nature's harvest by destroying biodiversity, increasing the use of herbicides and pesticides, and spreading the risk of irreversible genetic pollution.

According to the president of Monsanto, Hendrik Verfaillie, all biodiverse species that are not patented and owned by them are weeds that "steal the sunshine." Yet corporations that promote genetic engineering steal nature's harvest of diverse species, either by deliberately destroying biodiversity or by unintended biological pollution of species and ecosystems. They steal the global harvest of healthy and nutritious food. Finally, they steal knowledge from citizens by stifling independent science and denying consumers the right to know what is in their food.

---

From Vandana Shiva, *Stolen Harvest* (Cambridge, Mass.: South End Press, 2000). Reprinted by permission of the publisher.

## "Feeding the World"

"Feeding the world" is the main slogan of the biotechnology industry. In a $1.6 million European media blitz in 1998, Monsanto ran the following advertisement:

> Worrying About Starving Future Generations Won't Feed Them. Food Biotechnology Will.
>
> The world's population is growing rapidly, adding the equivalent of a China to the globe every 10 years. To feed these billion more mouths, we can try extending our farming land or squeezing greater harvests out of existing cultivation. With the planet set to double in numbers around 2030, this heavy dependency on land can only become heavier. Soil erosion and mineral depletion will exhaust the ground. Lands such as rainforests will be forced into cultivation. Fertilizer, insecticide, and herbicide use will increase globally.
>
> At Monsanto, we now believe food biotechnology is a better way forward. Our biotech seeds have naturally occurring beneficial genes inserted into their genetic structure to produce, say, insect- or pest-resistant crops.
>
> The implications for the sustainable development of food production are massive: Less chemical use in farming, saving scarce resources. More productive yields. Disease-resistant crops. While we'd never claim to have solved world hunger at a stroke, biotechnology provides one means to feed the world more effectively.
>
> Of course, we are primarily a business. We aim to make profits, acknowledging that there are other views of biotechnology than ours. That said, 20 government regulatory agencies around the world have approved crops grown from our seeds as safe.[1]

Hoechst, another self-styled "life sciences corporation," ran a similar ad in the April 16, 1999, *Financial Times*, asking us to "imagine a world where harvests grew just as fast as the population."

Ironically, Monsanto earns most of its revenue from the sale of chemicals, giving the lie to its claim that it is a "life sciences" company.[2] It attempts to cloak this fact by describing its sales of agrichemicals such as Roundup and related products as "agricultural" products rather than chemicals.

## cturing the Illusion of Sustainability

The "open" image that genetically engineered crops are sustainable is an illusion manufactured by corporations.

This illusion is created by several means. First, corporations attempt to portray biotechnology as an "information" technology with no material ecological impacts. As Monsanto's president has stated, "At the most basic level, then, biotechnology gives us the chance to achieve sustainability, by substituting information for stuff." What could be an easier god-trick than the argument that biotechnology achieves sustainability by "substituting information for stuff"? The material effects of genetic engineering disappear, and with them, the problem of negative ecological impacts. However, Roundup is "stuff," not information. Roundup Ready soybeans are stuff, Bollgard cotton is stuff, the genes engineered into it are stuff, and this stuff has ecological impact.

Second, corporations promote the misinformation that transgenic crops require fewer agrichemicals. In fact, evidence shows that transgenic crops lead to increased use of hazardous chemicals (see below).

Third, when corporations describe the benefits of genetic engineering, they do so in comparison to large-scale industrial agriculture rather than to ecological, small-scale agriculture. Yet most of the world's farmers are small-scale farmers working on less than two acres, both to meet their diverse food needs and to market some of their produce.

Biotech industry consultant Clive James claims that herbicide-resistant potatoes, for instance, save farmers $6 per acre, but this is based on a farm that spends between $30 and $120 per acre on insecticide control.[3] For an organic, ecological farm, herbicide-resistant potatoes increase costs by $25 to $115 per acre, and also require increased insecticide use.

## The Myth of Decreased Agrichemical Use

The development of herbicide-resistant and pest-resistant crops accounts for more than 80 percent of the biotechnology research in agriculture. However, evidence is already available that rather than controlling weeds, pests, and diseases, genetic engineering increases chemical use and can create superweeds, superpests, and superviruses.

Herbicide-resistance accounts for 71 percent of the applications of genetic engineering. Through genetically engineering herbicide resistance into crops, corporations are increasing sales of both chemicals and seeds. Monsanto's Roundup Ready soybeans are an example of such an herbicide-resistant crop.

The Roundup herbicide is Monsanto's flagship agricultural product. According to the company, Roundup, a glyphosate-based herbicide, "destroys every weed, everywhere, economically." However, Roundup is a non-selective herbicide that does not distinguish between weeds and desirable vegetation, and thus kills all plants, which is in no way economical. Roundup effectively controls a broad range of grasses and broadleaf weeds by inhibiting EPSP synthase, an enzyme essential to a plant's growth, and establishing a roadblock in the plant's metabolic pathways.

According to Monsanto,

> Many of you have heard of Monsanto's Roundup herbicide. And it's very effective at killing weeds—so effective, in fact, that Roundup would control soybeans as well as weeds if it should come into contact with both.
>
> At least, that was the case until Monsanto developed Roundup Ready Soybeans. Roundup Ready Soybeans express a novel protein that allows them to thrive, even when sprayed with enough Roundup to control competing weeds.[4]

The gene inserted in Roundup Ready crops increases the amount of EPSP synthase protein in the plants, providing a detour around Roundup's roadblock. Thus, in order to prevent weeds, farmers are encouraged to grow crops they do not necessarily need or consume.

In 1995, Monsanto genetically engineered a cotton plant, named Bollgard, meant to be resistant to the common bollworm pest. This transgenic crop is meant to enable farmers to dispense with the synthetic insecticides now used to control insect pests. However, the company admits that bollworm larvae more than one quarter inch long or older than two to four days are difficult to control with Bollgard alone.[5] According to Monsanto, "if sufficient larvae of this size are present you may need to apply supplemental treatment at intervals."[6]

The company suggests maintaining a refuge for Bollgard cotton: that is, it suggests that four acres of non-Bollgard cotton crops be planted as refuge for every 100 acres of Bollgard cotton planted. In India, the small-scale farmers that dominate the cotton-growing zones would find it very difficult to maintain such refuges.

In 1997, 20 percent of the first commercial crop of Roundup Ready cotton suffered deformed bolls and bolls dropping off early. During 1998, Monsanto started field trials of Bollgard in India with the aim of marketing genetically engineered seeds by 1999–2000. A review of pesticide sprays by the farmers at various trial sites in India revealed that the use of pesticides had not stopped at all for the Bollgard crop.[7]

Experiments with some caterpillar pests of cotton have proved that some pests (for example, *Spodoptera* and *Heliothis*) can develop resistance to the toxins engineered into Bollgard. Finally, since most crops have a diversity of insect pests, insecticides may still have to be applied to transgenic crops engineered to withstand just one pest. According to an analysis by the Pesticides Trust on behalf of Greenpeace, such herbicide-resistant varieties will alter the pattern of herbicide use, but will not change the overall amounts used.[8]

## The Myth of Increased Yields and Returns

Human ingenuity has always kept harvests above population growth. As Clifford Geertz has shown by comparing 22 farming systems, biodiversity and labor intensification are the most efficient and sustainable ways of increasing yields.

As Marc Lappé and Britt Bailey report in their book *Against the Grain*, herbicide-resistant soybeans yielded 36 to 38 bushels per acre, while hand-tilled soybeans yielded 38.2 bushels per acre. According to the authors, this raises the possibility that the gene inserted into these engineered plants may selectively disadvantage their growth when herbicides are not applied. "If true, data such as these cast doubt on Monsanto's principal point that their genetic engineering is both botanically and environmentally neutral," the authors write.[9]

In any case, in the corporate-controlled food system, the same company may perform the research, sell the seeds, and provide the data about its products. Thus, the patient, diagnostician, and physician are rolled into one, and there is no objective basis of assessment of yield performance or ecological impact.

Although Monsanto's Indian advertising campaign reports a 50 percent increase in yields for its Bollgard cotton, a survey conducted by the Research Foundation for Science, Technology, and Ecology found that the yields in all trial plots were lower than what the company promised. Yields from the local, cultivated hybrid variety and Bollgard were more or less the same.

Bollgard's failure to deliver higher yields has been reported all over the world. The Mississippi Seed Arbitration Council ruled that in 1997, Monsanto's Roundup Ready cotton failed to perform as advertised, recommending payments of nearly $2 million to three cotton farmers who suffered severe crop losses.

While increased food productivity is the argument used to promote genetic engineering, when the issue of potential adverse impacts on farmers is brought up, the biotechnology industry itself argues that genetic engineering does *not* lead to increased productivity. Thus Robert Shapiro, CEO of Monsanto, while referring to Posilac (Monsanto's bovine growth hormone) in *Business Ethics*, said on the one hand that

> There is need for agricultural productivity, including dairy productivity, to double if we want to feed all the people who will be joining us, so I think this is unequivocally a good product.[10]

On the other hand, when asked about the product's economic impact on farmers, he said that it would "play a relatively small role in the process of increasing dairy productivity."

## The Socioeconomic Costs of Genetically Engineered Seeds

Cultivating genetically modified crops is more expensive than conventional crops because of the higher costs of the seed, technology fees, and the need for increased use of chemicals. In organic agriculture, the seeds are saved and cultivated the following season, and other necessary inputs for the seeds' cultivation are provided on the farm. When genetically engineered seeds are cultivated, all of these inputs must be paid for, and farmers will inevitably encounter serious financial troubles. Cultivating Bollgard cotton is estimated to cost Indian farmers nearly nine times more than cultivating a conventional variety. If the 21.4 million acres under cotton cultivation in India in 1997–98 were shifted to genetically engineered cotton, it would cost nearly Rs. 224.7 billion.

These increased costs can push farmers into bankruptcy and even suicide. The 1998 failure of the hybrid cotton crop in Andhra Pradesh due to pest devastation, and the subsequent suicide of farmers due to indebtedness—caused by spending nearly Rs. 12,000 per acre on pesticides—indicate how vulnerable our agricultural systems have become.

## th of Safe Foods

and other corporations repeatedly refer to their seeds and foods as having been tested for safety. But not only have no ecological or food-safety tests been conducted on genetically engineered crops and foods before commercialization; corporations have tried every means within their reach to steal the right to safe and nutritious food from citizens and consumers.

It is often claimed that there have been no adverse consequences from over 500 field releases in the United States. In 1993, for the first time, the data from the U.S. Department of Agriculture (USDA) field trials were evaluated to see whether they support these safety claims. The Union of Concerned Scientists (UCS), which conducted the evaluation, found that the data collected by the USDA on small-scale tests have little value for commercial risk-assessment. Many reports fail to even mention—much less measure—environmental risks. Of those reports that allude to environmental risk, most have only visually scanned field plots looking for stray plants or isolated test crops from relatives. The UCS concluded that the observations that "nothing happened" in those hundreds of tests do not say much. In many cases, adverse impacts are subtle and would never be registered by scanning a field. In other cases, failure to observe evidence of the risk is due to the contained conditions of the tests. Many test crops are routinely isolated from wild relatives, a situation that guarantees no out-crossing. The UCS cautioned that "care should be taken in citing the field test record as strong evidence for the safety of genetically engineered crops."[11]

All genetically engineered crops use genes that are resistant to antibiotics to help identify whether the genes that have been introduced from other organisms have been successfully inserted into the engineered crop. These marker genes can exacerbate the spread of antibiotic resistance among humans. Based on this concern, Britain rejected Ciba-Geigy's transgenic maize, which contains the weaker gene for campicillin resistance.

Many transgenic plants are engineered for resistance to viral diseases by incorporating the gene for the virus's coat protein. These viral genes may cause new diseases. New broad-range recombinant viruses could arise, causing major epidemics.

Upon consumption, the genetically engineered DNA of these foods can break down and enter the bloodstream. It has long been assumed that the human gut is full of enzymes that can rapidly digest DNA. But in a study designed to test the survival of viral DNA in the gut, mice were fed DNA from a bacterial virus, and large fragments were found to survive passage through the gut and to enter the bloodstream.[12] Further studies indicate that the ingested DNA can end up in the spleen and liver cells as well as in white blood cells.[13]

Within the gut, vectors carrying antibiotic-resistance markers may also be taken up by the gut bacteria, which would then serve as a mobile reservoir of antibiotic-resistance genes for pathogenic bacteria. Horizontal gene transfer between gut bacteria has already been demonstrated in mice and chickens and in human beings.[14]

When L-tryptophan, a nutritional supplement, was genetically engineered and first marketed, 37 people died and 1,500 people were severely affected by a painful and debilitating circulatory disorder called eosinophilia myalgia.[15] When a gene from the Brazil nut was inserted into soybeans to increase their protein levels, the transgenic soybeans also contained the nut's allergenic properties.[16]

Greenpeace and other non-governmental organizations have revealed that soybean plants sprayed with Roundup are more estrogenic and could act as hormone or endocrine-system disrupters. Dairy cows that consume Roundup Ready soybeans produce milk with higher fat levels than cows that eat regular soybeans.

## The Myth of Food Security

The Green Revolution narrowed the basis of food security by displacing diverse nutritious food grains and spreading monocultures of rice, wheat, and maize. However, the Green Revolution focused on staple foods and their yields. The genetic engineering revolution is undoing the narrow gains of the Green Revolution both by neglecting the diversity of staples and by focusing on herbicide resistance, not higher yields.

According to Clive James, transgenic crops are not engineered for higher yields. Fifty-four percent of the increase in transgenic crops is for those engineered for herbicide resistance, or, rather, the increased use of herbicides, not increased food. As an industry briefing paper states, "The herbicide tolerant gene has no effect on yield per se."[17] Worldwide, 40 percent of the land under cultivation by genetically engineered crops is under soybean cultivation, 25 percent under corn, 13 percent under tobacco, 11 percent under cotton, 10 percent under canola, and 1 percent each under tomato and potato. Tobacco and cotton are non-food commercial crops, and crops such as soybeans have not been food staples for most cultures outside East Asia. Such crops will not feed the hungry. Soybeans will not provide food security for *dal*-eating Indians, and corn will not provide security in the sorghum belt of Africa.

The trend toward the cultivation of genetically engineered crops indicates a clear narrowing of the genetic basis of our food supply. Currently, there are only two commercialized staple-food crops. In place of hundreds of legumes and beans eaten around the world, there is soybean. In place of diverse varieties of millets, wheats, and rices, there is only corn. In place of the diversity of oil seeds, there is only canola.

These crops are based on expanding monocultures of the same variety engineered for a single function. In 1996, 1.9 million acres around the world were planted with only two varieties of transgenic cotton, and 1.3 million acres were planted with Roundup Ready soybeans. As the biotechnology industry globalizes, these monoculture tendencies will increase, thus further displacing agricultural biodiversity and creating ecological vulnerability.

Further, by forcing the expansion of non-food crops such as tobacco and cotton, transgenic crops result in fewer acres in food production, aggravating food insecurity.

## The Destruction of Biodiversity

In Indian agriculture, women use up to 150 different species of plants (which the biotech industry would call weeds) as medicine, food, or fodder. For the poorest, this biodiversity is the most important resource for survival. In West Bengal, 124 "weed" species collected from rice fields have economic importance for local farmers. In a Tanzanian village, over 80 percent of the vegetable dishes are prepared from uncultivated plants.[18] Herbicides such as Roundup and the transgenic crops engineered to withstand them therefore destroy the economies of the poorest, especially women. What is a weed for Monsanto is a medicinal plant or food for rural people.

Since biodiversity and polycultures are an important source of food for the rural poor, and since polycultures are the most effective means of soil conservation, water conservation, and ecological pest and weed control, the Roundup Ready technologies are in fact a direct assault on food security and ecological security.

## The Risks of Genetic Pollution

Genetically engineered crops increase chemical use and add new risks of genetic pollution. Herbicide-resistant crops are designed for intensive use of herbicides in agriculture. But they also create the risks of weeds being transformed into "superweeds" by the transfer of herbicide-resistant traits from the genetically engineered crops to closely related plants.

Research in Denmark has shown that oilseed rape genetically engineered to be herbicide-tolerant could transmit its introduced gene to a weedy natural relative through hybridization. Weedy relatives of rape are now common in Denmark and throughout the world. Converting these "weeds" into "superweeds" that carry the gene for herbicide-resistance would provoke high crop losses and increasing use of herbicides. For these reasons, the European Union has imposed a *de facto* moratorium on the commercial planting of genetically engineered crops.

In many cases, the weeds that plague cultivated crops are relatives of the crops themselves. Wild beets have been a major problem in European sugar-beet cultivation since the 1970s. Given the gene exchange between weedy beets and cultivated beets, herbicide-resistant sugar beets could only be a temporary solution.[19]

Superweeds could lead to "bioinvasions," displacing local diversity and taking over entire ecosystems. The problem of invasive species is being increasingly recognized as a major threat to biodiversity. Monsanto's claim that products such as Roundup Ready soybeans will reduce herbicide use is false because it does not take into account the introduction of such engineered plants in regions where herbicides are not used in agriculture and where native diversity of soybeans exists. China, Taiwan, Japan, and Korea are regions where soybeans have evolved and where wild relatives of cultivated soybeans are found. In these regions, Monsanto's Roundup Ready soybeans would increase herbicide use and "pollute" the native biodiversity by transferring herbicide-resistant genes to wild plants. This could lead to new weed problems and loss of biodiversity. Moreover, since the Third World is the home to most of the world's biodiversity, the risks of genetic pollution in Third World countries are even more profound.

Herbicide-resistant transgenic crops can also become weeds when seeds from those crops germinate after harvest. More herbicides will have to be applied to eliminate these "volunteer plants."

## Toxic Plants: A Recipe for Superpests

The bacterium *Bacillus thuringiensis* (Bt) was isolated from soil in 1911. Since 1930, it has been available as an organic form of pest control. Organic farmers have stepped up its use since the 1980s.

Monsanto and other "life sciences" corporations developed a technique of inserting the toxin-producing gene from the Bt bacteria into plants. This particular Bt gene produces a toxin that disables insects, and the genetically engineered Bt plants are thus able to produce their own pesticide. Genetically engineered Bt-crops have been cultivated commercially since 1996.

While Monsanto sells Bt-crops with the claim that they will reduce pesticide use, Bt-crops can actually create "superpests" and increase the need for pesticides. Bt-crops continuously express the Bt toxin throughout their growing season. Long-term exposure to the toxins promotes the development of resistance in insect populations. This kind of exposure could lead to selection for resistance in all stages of the insect pest on all parts of the plant for the entire season.

Due to these risks of encouraging pest resistance, the U.S. Environmental Protection Agency (EPA) offers only conditional and temporary registration for Bt-crops. The EPA requires a 4 percent refuge for Bt cotton—i.e., 4 percent of the cotton in a Bt-cotton field must be conventional and not express the Bt toxin. The conventional cotton acts as a refuge for insects to survive and breed in order to keep the overall level of resistance in the population low.

While the Monsanto propaganda states that farmers will not have to use pesticides, the reality is that the management of resistance requires continued use of non-Bt cotton and pesticide sprays. And even with a 4 percent refuge, insect resistance will evolve in as few as three to four years. Already eight species of insects have developed resistance to Bt toxins, including diamond black moth, Indian meal moth, tobacco budworm, colorado potato beetle, and two species of mosquitoes.[20]

Even if Bt-crops do repel some pests, most crops have a diversity of insect pests. Insecticides will still have to be applied to control pests that are not susceptible to Bt's toxin. Beneficial species such as birds, bees, butterflies, and beetles, which are necessary for pollination and which through the prey-predator balance also control pests, may be threatened by Bt-crops.[21] Soil-inhabiting organisms that degrade the toxin-contaminated organic matter can be harmed by the toxin. Nothing is known of the impact on human health when Bt-crops such as potato and corn are eaten, or on animal health when oilcake from Bt-cotton or fodder from Bt-corn is consumed as cattle feed.

## The Politics of Biosafety

Biosafety, or the prevention of biohazards caused by genetic engineering, is emerging as the most important environmental and scientific issue of our time. Biosafety issues are intimately linked to the politics of science, and to the conflicting perspectives of different scientific cultures and traditions.

One conflict is between the ecological sciences that assess the impact of genetic engineering on the environment and on human health, and reductionist sciences that promote production based on genetic engineering.

A second conflict is between private-interest and public-interest science. When the techniques of recombinant DNA were emerging in the late 1970s and 1980s, the crippled organisms that resulted from the experiments were not meant to survive in the environment. The main practitioners during this phase were university scientists, and they themselves called for a moratorium on recombinant DNA research.

During the 1980s and 1990s, scientists who had developed genetic engineering techniques left universities to start biotechnology firms. During this phase, concerns for safety were sidelined by the promise of biotech miracles. Today, genetically engineered organisms are being released for production and consumption on global markets, and small, start-up biotech firms are being bought up by giant chemical corporations.

The biosafety issues that were outlined by university scientists using crippled organisms are very different from those posed by robust organisms being produced by transnational corporations for global markets. These issues interfere in the market expansion of genetic engineering in agriculture, and thus industry has attempted to suppress the debate in four main ways.

First, they invoke a call to "sound science," which they equate with industry-friendly science, and treat industry-independent science as "junk science." "Sound science" has become like a mantra for banishing safety regulations. This was the phrase used by the industry in a letter to President Clinton at the G7 Summit in Denver in 1997.[22] It is the language of *The Wall Street Journal* editorial accusing Europe of practicing "junk science" by banning the import of hormone-fed beef, and referring to the World Trade Organization (WTO) decision against the ban as "real science."[23] According to the U.S. agricultural secretary, Dan Glickman, who has stated categorically that the United States will stand behind its genetically engineered foods and oppose any European labeling requirements as a trade violation,

> We've got to make sure that sound science prevails, not what I call historic culture, which is not based on sound science. Europe has a much greater sensitivity to the culture of food as opposed to the science of food. But in the modern world, we just have to keep

the pressure on the science. Good science must prevail in these decisions.[24]

However, the conflict over genetically engineered crops and foods is not a conflict between "culture" and "science." It is between two cultures of science: one based on transparency, public accountability, and responsibility toward the environment and people, and another based on profits and the lack of transparency, accountability, and responsibility.

Second, the industry claims that there is "substantial equivalence" between genetically engineered products and natural ones. When corporations claim monopoly rights to seeds and crops, they refer to genetically modified organisms (GMOs) as "novel." When the same corporations want to disown risks by stifling safety assessment and analysis of hazards, they refer to transgenic organisms as being substantially equivalent to their naturally occurring counterparts. the same organism cannot be both "novel" and "not novel." This ontological schizophrenia is a convenient construct to create a regime of absolute rights and absolute irresponsibility. Through the WTO, the ontological schizophrenia is being spread from the United States to the rest of the world.

The genetic engineering guidelines of the Food and Drug Administration (FDA) are based on the assumption that GMOs behave like their naturally occurring counterparts. The guidelines are also based on the assumption that "genetically engineered organisms have greater predictability compared to species evolved by traditional techniques." Neither of these assumptions is true. GMOs do not behave like their naturally occurring counterparts, and the behavior of GMOs is highly unpredictable and unstable.

For example, naturally occurring *Klepsiella planticola* does not kill plants, but, as research at the University of Oregon has shown, the genetically engineered *Klepsiella* was lethal to crops.[25] The naturally occurring *Bacillus thuringiensis* has not contributed to the evolution of resistance in pests, but the genetically engineered Bt-crops create rapid resistance evolution because the Bt toxin is expressed in *every* cell of the plant, *all* the time. Thus the assumption of "substantial equivalence" does not hold.

The assumption of "predictability" is also totally false. While genetic engineering makes the *identification* of the gene to be transferred into another organism more predictable, the ecological *behavior* of the transferred gene in the host genome is totally unpredictable. A transgenic yeast, which was engineered to ferment faster, accumulated a certain metabolite at toxic levels. Between 64 and 92 percent of the first generation of transgenic tobacco plants is unstable. Petunias do not have unstable coloring, but genetically engineered petunias change their color unpredictably due to "gene silencing."[26]

In 1998, when Dr. Arpad Pusztai concluded from experiments on rats that there was a lack of equivalence in both composition and metabolic consequences between genetically engineered and conventional potatoes, he was sacrificed to protect corporate control and profits. Pusztai was suspended by his lab, accused of scientific fraud, and banned from speaking to the media about his results. In 1999, 20 scientists from 14 countries examined the Pusztai report and accused his employer, the Rowett Institute in Scotland, of bowing to public pressure. Claims of a cover-up were reinforced when it was revealed that Rowett had received £140,000 of funding from Monsanto. In 1999, Dr. S. W. B. Even, a senior pathologist at the University of Aberdeen, provided conclusive evidence supporting Pusztai's findings.[27]

Third, as has been discussed above, the biotech industry further attempts to elide biosafety issues by describing contained, artificially constructed experiments as "field trials" that prove safety, and by arguing that the labeling of genetically engineered foods, guaranteeing consumers the "right to know" and the "right to choose," interferes with free trade.

Fourth and finally, the ultimate step in total control over the food system is the attempt by the USDA to destroy the organic option for farmers and consumers. If adopted and implemented, the USDA policy would outlaw genuine organic production all over the world.

Under this policy, the USDA will allow fruit and vegetables that have been genetically engineered, irradiated, treated with additives, and raised on contaminated sewage sludge to be labeled "organic." "Organic" livestock can be housed in batteries, fed with the offal of other animals, and injected with biotics.

Further, the policy prohibits the setting of any standards higher than those established by the department. Farmers will, in other words, be forbidden by law from producing and selling good, safe food. As Thames University professor George Monbiot writes, "Organic produce, in the brave new world of American oligopoly, will be virtually undistinguishable from conventionally toxic food."[28] To date, the policy has been stalled by virtue of a major citizen mobilization against it.

## The Subversion of Biosafety Laws

The United Nations Convention on Biological Diversity (CBD) outlined international biosafety laws. A small team from the Third World Network worked closely with Third World governments to introduce these rules into the CBD. Article 19.3 of the Convention states,

> the Parties shall consider the need for . . . appropriate procedures, including, in particular, advance informed agreement, in the field of the safe transfer, handling, and use of any living modified organism resulting from biotechnology that may have adverse effect on the conservation and sustainable use of biological diversity.

The language of "living modified organism" was introduced by the United States in place of "genetically modified organism" to neutralize public concern about genetic engineering. "Living modified organism" applies to all products of conventional breeding, not just genetically engineered species. Then-President George Bush refused to sign the CBD because, according to him, it would interfere with the growth of the $50 billion U.S. biotech industry.

In spite of not being a party to the CBD, the United States has been present at every negotiation regarding the convention. It tried to undo the work of Panel IV, set up by the United Nations to implement CBD articles on biosafety. Although environmentalists succeeded in keeping the issue of biosafety alive for seven years despite U.S. intransigence and irrationality, a small group of countries including the United States killed the Biosafety Protocol in 1999, on the grounds that it would interfere with WTO free-trade rules.

## Cultivating Diversity

In the mountain farming systems of the Garhwal Himalaya, there is a particular cropping pattern called *baranaja*, which means literally "12 seeds." The seeds of 12 or more different crops are mixed and then randomly sown in a field fertilized with cow dung and farmyard manure. Care is taken to balance the distribution of the crops in each area of the field. After sowing, the farmer transplants crops from one area of the field to another area in order to maintain an even distribution of the crops. As in other cultivation practices, constant weeding is necessary. The crops are all sown in May, but are harvested at different times, from late August to early November, thus ensuring a continuous food supply for the farmer during this period and beyond. The different crops have been selected by the farmers over the ages by observing certain relationships between plants, and between plants and soil. For example, the *rajma* creeper will climb only on the *marsha* plant and on no other plant in the field.

The symbiotic relationships between different plants contribute to the increased productivity of the crops. When farmers cultivate *baranaja*, they get higher yields, diverse outputs, and a better market price for their produce than when they cultivate a monoculture of soybeans. Soybeans sell for only Rs. 5 per kilogram, whereas *jakhia*, one of the *baranaja* crops that matures the earliest, sells for Rs. 60 per kilogram.

Cultivating diversity can therefore be part of a farming strategy for high yields and high incomes. But since these yields and incomes are from diverse crops, centralized commercial interests are not interested in them. For them, uniformity and monocultures are an imperative. However, from the point of view of small farmers, diversity is both highly productive and sustainable.[29]

## Genetic Engineering and Food Security

Diversity and high productivity go hand in hand if diverse outputs are taken into account and the costs of external inputs are added to the cost of inputs. The monoculture paradigm focuses on yields of single commodities and externalizes the costs of chemicals and energy. Inefficient and wasteful industrial agriculture are hence presented as efficient and productive.[30]

The myth of increasing yields is the most common justification for introducing genetically engineered crops in agriculture. However, genetic engineering is actually leading to a "yield drag." On the basis of 8,200 university-based soybean trials in 1998, it was found that the top Roundup Ready soybean varieties had 4.6 bushels per acre, or yields 6.7 percent lower than the top conventional varieties. As environmental consultant Dr. Charles Benbrook states,

> In 1999, the Roundup Ready Soybean yield drag could result in perhaps a 2.0 to 2.5 percent reduction in national average soybean yields, compared to what they would have been if seed companies had not dramatically shifted breeding priorities to focus on herbicide tolerance. If not reversed by future breeding enhancements, this downward shift in soybean yield potential could emerge as the most significant decline in a major crop ever associated with a single genetic modification.[31]

Research on trials with Bt cotton in India also showed a dramatic reduction in yields: in some cases as high as 75 percent.[32]

As criticism of biotechnology's emphasis on herbicide-resistant crops and crops that produce toxins grows, the biotechnology industry has started to talk of engineering crops for nitrogen fixing, salinity tolerance, and high nutrition instead. However, all these traits already exist in farmers' varieties and farmers' fields. Legumes and pulses intercropped with cereals fix nitrogen. In coastal ecosystems, farmers have evolved a variety of salt-tolerant crops. We do not need genetic engineering to give us crops rich in nutrition. Amaranth has nine times more calcium than wheat and 40 times more calcium than rice. Its iron content is four times higher than that of rice, and it has twice as much protein. *Ragi* (finger millet) provides 35 times more calcium than rice, twice as much iron, and five times more minerals. Barnyard millet contains nine times more minerals than rice. Nutritious and resource-prudent crops such as millets and legumes are the best path of food security.

Biodiversity already holds the answers to many of the problems for which genetic engineering is being offered as a solution. Shifting from the monoculture mind to biodiversity, from the engineering paradigm to an ecological one, can help us conserve biodiversity, meet our needs for food and nutrition, and avoid the risks of genetic pollution.

## Notes

1. "Monsanto: Peddling 'Life Sciences' or 'Death Sciences'?" New Delhi: Research Foundation for Science, Technology, and Ecology (RFSTE), 1998.

2. "Monsanto: Peddling 'Life Sciences' or 'Death Sciences'?" p. 12.

3. Clive James, "Global Status of Transgenic Crops in 1997," *ISAAA Briefs*, 1997, p. 20.

4. International Association of Plant Breeders, "Feeding the 8 Billion and Preserving the Planet," NYON, Switzerland.

5. Monsanto promotional material, 1996.

6. Monsanto, Bollgard, 1996.

7. Vandana Shiva, Afsar Jafri, and Ashok Emani, "Globalization of the Seed Sector," Bombay: EPW, 1999.

8. International Agricultural Development, 1998.

9. Marc Lappé and Britt Bailey, *Against the Grain: Biotechnology and the Corporate Takeover of Your Food*, Monroe, ME: Common Courage Press, 1998.

10. Interview with Robert Shapiro, *Business Ethics*, January-February 1996, p. 47.

11. Margaret Mellon and Jane Rissler, *Risks of Genetically Engineered Crops*, Cambridge, MA: MIT Press, 1996.

12. Mae-Wan Ho, *Genetic Engineering: Dream or Nightmare*, Bath, U.K.: Gateway Books, 1998, p. 165.

13. Philip Cohen, "Can DNA in food find its way into cells?" *New Scientist*, January 4, 1997, p. 14.

14. Mae-Wan Ho.

15. Lappé and Bailey, p. 134.

16. J. A. Nordlee et al., "Identification of a Brazil Nut Allergen in Transgenic Soybeans," *The New England Journal of Medicine*, No. 334, 1996, pp. 688–92.

17. Clive James, p. 14.

18. Jane Rissler and Margaret Mellon, *The Ecological Risks of Engineered Crops*, Cambridge, MA: MIT Press, 1996.

19. P. Bondry, M. Morchen, et al., "The origin and evolution of weed beets: consequences for the breeding and release of herbicide resistant transgenic sugar beets," *Theoretical and Applied Genetics*, No. 87, 1993, pp. 471–78.

20. Miguel Altieri, "Ecological Impact of Genetic Engineering," unpublished paper, 1998.

21. Vandana Shiva and Afsar H. Jafri, "Seeds of Suicide," RFSTE, 1998.

22. Letter of U.S. Agribusiness to President Clinton at G7 Summit, Denver, June 18, 1997.

23. *Wall Street Journal* Editorial, November 6, 1997.

24. Dan Glickman, quoted in Vandana Shiva, *Betting on Biodiversity*, New Delhi: RFSTE, 1998, p. 45.

25. Report of the Independent Group of Scientific and Legal Experts on Biosafety, 1996.

26. Report of the Independent Group, 1996.

27. COST 98 Action (European Union Program) in Lund, Sweden, November 25–27, 1998.

28. George Monbiot, "Food Fascism," *Guardian*, March 3, 1998.

29. Research Foundation for Science, Technology, and Natural Resource Policy, "Cultivating Diversity: Biodiversity Conservation and the Politics of the Seed," New Delhi, 1993.

30. Vandana Shiva, "Biodiversity-Based Productivity," New Delhi: RFSTE, 1998; and Peter Rosset and Miguel Altieri, "The Multiple Functions and Benefits of Small Farm Agriculture," International Forum on Agriculture, San Francisco, 1999.

31. Charles Benbrook, "Evidence of the Magnitude and Consequences of the Roundup Ready Soybean Yield Drag from University-Based Varietal Trials in 1998," InfoNet Technical Paper, No. 1, Sandpoint, Ohio: July 13, 1999, p. 1.

32. Vandana Shiva et al., "Globalization and Seed Security: Transgenic Cotton Trials," *EPW*, Vol. 34, No. 10–11, March 6–19, 1999, p. 605.

# GM Food Is the Best Option We Have        14

ANTHONY J. TREWAVAS

*Anthony Trewavas is a professor at the Institute of Cell and Molecular Biology at the University of Edinburgh in Scotland. He earned a Ph.D. in biochemistry from University College, London, in 1964 and was a postdoctoral fellow in the School of Biological Sciences at the University of East Anglia from 1964 to 1970. He joined the University of Edinburgh in 1970, where he has taught and done research for thirty years.*

*Trewavas's lab is called the Edinburgh Molecular Signalling Group. The team that he leads there focuses on the role of calcium in signal transduction during plant growth as well as the role of calcium in all aspects of plant development.*

*Trewavas is known as a defender of biotechnology and GM food in England, where it is not easy to champion such a view. In the essay presented here, Trewavas argues that Golden Rice is a wonderful opportunity to help the poor of the world and that we must embrace scientific advances in plant technology, not fear them. Using rapeseed and other examples, he dismisses fears about "superweeds" and "Frankenfoods," and he argues that GM plants hold the key to a better agricultural future for everyone.*

I HAVE BEEN A plant biologist for 40 years. What drew me to the subject was love of the organism. All those that deal with plants will know this feeling of pleasure and peace that comes from contact. We use plants in many different ways; for food, clothes, timber, cooking and drugs and to beautify our environment. To improve these uses for human benefit we must first gain better understanding of the way such complex organisms work. My respect has grown the more I have come to understand the beautiful and intricate way in which plants function. Our role on this planet is to act the

---

From Anthony Trewavas, "GM Is the Best Option We Have," unpublished manuscript, which appeared on the now-defunct AgBioView listserv (http://agbioview.listbot.com). Reprinted by permission of the author.

good gardener. Like all such gardeners or stewards we seek to provide a planetary garden which survives in harmony with itself. But this garden can only be in harmony when all our fellow men and women, the other stewards of this planet, can enjoy a complete and fulfilling life enabling the full flowering of the potential in all of us.

There are some people in this country that stereotype scientists without ever knowing any of them; that ascribe ulterior motives to scientific endeavour and surround themselves with acolytes of similar limited experience. These people commonly rate the wisdom of nature as superior to human ingenuity and survival.

But we investigate nature so that we can stop the natural things that destroy our lives and curtail our stewardship. I am talking about natural things like child death, leprosy, disease-ridden water, starvation or floods that are clearly part of nature and nature's wisdom. Human ingenuity, which our opponents cast so easily aside, has given us antibiotics, anaesthetics and warm houses to prolong and protect life, security of food supply, transport to places so that we can share in the pride and glory of human achievements in arts, music and architecture; and has even taken us to the moon. All these ingenuities derive from knowledge of the world in which we live and result from experimentation and improvement of nature, the "good gardening" which opponents denigrate. There is a desire by some to reverse history, to recover some mythical golden age when life expectancy was under half what it is now; when people died needlessly and painfully from a variety of unknown causes (some most certainly from diseases in their poor quality food) and when, for example in the UK, half the young men called up for the Boer war were refused on the grounds of serious underweight, height and poor health identified as resulting from malnourishment. When problems develop we must continue to rise to the challenge to tackle them as we have done in the past with nobility and intellect. Do not listen to the siren voices that say "stop the world I want to get off". There are many such voices in the UK at present.

A decade ago, as a university plant biologist, I thought that genetic manipulation GM would be publicly funded and used for the benefit of mankind. Indeed I share in the general distrust of GM commercialisation and I know this is a major complication in the UK. But this is the world we live in; if you don't like it change the economics, don't demean the knowledge. We can't eliminate knowledge simply because someone makes a profit out of it.

Two recent reports of publicly funded, university GM research now indicate its true potential. US scientists in collaboration with Japanese workers have genetically improved (GM) rice to increase seed yield of each plant by 35%. Why is this important?

One of the most certain facts about the human population is that it is increasing. By 2025 there will be 2.3 billion extra souls on mother earth, 50 times the current population of the UK, and they will have to be fed. Our current numbers of some six billion have already placed dangerous burdens on the ecosystems of spaceship earth and threaten our bio-diversity on which we are all interdependent. Global warming may indeed be global warning. So ploughing up wilderness to feed these extra people is no option. We can also eliminate organic farming as a meaningful solution. Organic farmers rely ultimately and only on soil nitrogen fixation to provide the essential nitrate and ammonia for crop growth and yield. Rainwater provides the other minerals. Since the maximum yields of fixed nitrogen have been measured numerous times we can estimate that by taking another 750 million ha [1 hectare = 10,000 sq. meters or 2.471 acres—*Ed.*] of wilderness under the plough we could feed just three billion. When Greenpeace tells us to "go organic" I ask myself which three billion will live and which three billion will die; perhaps they can enlighten us when they have finished tangling with the courts.

Clever plant breeding in the early 60's produced rice and wheat plants with well over double their previous yield; such progress enabled a parallel doubling of mankind, without massive starvation. But this option is now exhausted. Ignoring the problem, leaving billions to starve in misery, the worst of all tortures according to Amnesty International, is not an option either. "Every man's death diminishes me because I am part of mankind; ask not for whom the bell tolls . . ." is a philosophy I know many here will share with John Donne. So where one grain grew before we now again have to ensure that two will grow in the future. Currently GM is our best option to achieve this difficult task. This first report is very encouraging.

Critics say to me there is enough food to feed the world and they may well be right, at present. We produce sufficient to feed 6.4 billion people but the excess is largely in the West and it is far easier for scientists to conjure more food from the plants we grow than to persuade the West to share its agricultural bounty with its poorer neighbours. But the excess will not last long; our population increases by 1.3%/year, current annual cereal increases are only 1.1%. We live on the residual excess produced by the green revolution. At some point catastrophe beckons.

Our second report deals with a problem that kills one million young children in the third world every year and leaves many millions permanently blind. For a variety of reasons, babies can be prematurely weaned off breast milk. It's not a problem in the West, a variety of other foods and milk are available. But in the backwoods of the Far East, the usual option is rice gruel. Rice however contains no vitamin A and such babies rapidly become deficient. Either eye development is permanently damaged (we all need vitamin A for sight), or they succumb to childhood diseases that any western baby shrugs off in a week. Scientists in a Swiss university in a "tour de force" have genetically improved rice to make vitamin A. This golden rice has been given to the International Rice Institute in the Philippines for distribution to help ameliorate this serious problem and ensure a better life for parents and children.

There are some Western critics who oppose any solution to world problems involving technological progress. They denigrate this remarkable achievement. These luddite individuals found in some Aid organisations instead attempt to impose their primitivist western views on those countries where blindness and child death are common. This new form of Western cultural domination or neo-colonialism, because such it is, should be repelled by all those of good will. Those who stand to benefit in the third world will then be enabled to make their own choice freely about what they want for their own children.

But these are foreign examples; global warming is the problem that requires the UK to develop GM technology. 1998 was the warmest year in the last one thousand years. Many think global warming will simply lead to a wetter climate and be benign. I do not. Excess rainfall in northern seas has been predicted to halt the Gulf Stream. In this situation, average UK temperatures would fall by 5 degrees centigrade and give us Moscow-like winters. There are already worrying signs of salinity changes in the deep oceans. Agriculture would be seriously damaged and necessitate the rapid development of new crop varieties to secure our food supply. We would not have much warning. Recent detailed analysis of arctic ice cores has shown that the climate can switch between stable states in fractions of a decade. Even if the climate is only wetter and warmer new crop pests and rampant disease will be the consequence. GM technology can enable new crops to be constructed in months and to be in the fields within a few years. This is the unique benefit GM offers. The UK populace needs to be much more positive about GM or we may pay a very heavy price.

In 535 A.D. a volcano near the present Krakatoa exploded with the force of 200 million Hiroshima A bombs. The dense cloud of dust so reduced the intensity of the sun that for at least two years thereafter, summer turned to winter and crops here and elsewhere in the Northern hemisphere failed completely. The population survived by hunting a rapidly vanishing population of edible animals. The after-effects continued for a decade and human history was changed irreversibly. But the planet recovered. Such examples of benign nature's wisdom, in full flood as it were, dwarf and make miniscule the tiny modifications we make upon our environment. There are apparently 100 such volcanoes round the world that could at any time unleash forces as great. And even smaller volcanic explosions change our climate and can easily threaten the security of our food supply. Our hold on this planet is tenuous. In the present day an equivalent 535 A.D. explosion would destroy much of our civilisation. Only those with agricultural technology sufficiently advanced would have a chance at survival. Colliding asteroids are another problem that requires us to be forward-looking accepting that technological advance may be the only buffer between us and annihilation.

When people say to me they do not need GM, I am astonished at their prescience, their ability to read a benign future in a crystal ball that I cannot. Now is the time to experiment; not when a holocaust is upon us and it is too late. GM is a technology whose time has come and just in the nick of time. With each billion that mankind has added to the planet have come technological advances to increase food supply. In the 18th century, the start of agricultural mechanisation; in the 19th century, knowledge of crop mineral requirements, the eventual Haber Bosch process for nitrogen reduction. In the 20th century plant genetics and breeding, and later the green revolution. Each time population growth has been sustained without enormous loss of life through starvation even though crisis often beckoned. For the 21st century, genetic manipulation is our primary hope to maintain developing and complex technological civilisations. When the climate is changing in unpredictable ways, diversity in agricultural technology is a strength and a necessity not a luxury. Diversity helps secure our food supply. We have heard much of the precautionary principle in recent years; my version of it is "be prepared".

But how do these examples compare with the scepticism shown by the UK public over GM food; doesn't it harm human health? What about those apocalyptic visions of damage to the environment propounded by green organisations? If these views had any real substance I would share them, but they are totally contrary to all experience.

The testing of GM food is exemplary in its detail and takes at least four years. Sir John Krebs, head of our new Food Standards Agency, concluded that GM food is as safe as its non-GM counterpart. If eating foreign DNA and protein is dangerous we have been doing so for all of our lives with no apparent effects. Each GM food will be considered by regulatory authorities on its own merit.

As for GM environmental effects, many countries provide us with details of reduced use of herbicides and pesticides of 15–100%, of increased crop yields, less insect damage, a return of non-target insects to fields and reductions in fungal toxins in food. Even the flurry over the Monarch butterfly has been capped by record numbers on migration last year. Over 20 laboratories have now shown the original Monarch fears were groundless. Within five years, vaccines against the killer *E. coli*, hepatitis B, cholera and other diseases will all come in GM food. Even now they are in human trials. These vaccines will be very stable, be easily distributed world-wide, need no refrigeration or injection, merely consumption. The great campaign to eliminate world-wide disease, as we have with smallpox, will be well under way. Apocalypse now? Hardly.

Many of you may think that environmentalists are synonymous with ecologists. You would be mistaken. Let me read out for you extracts from what has become known as the Aachen declaration made by a large number of ecologists.

"Today's campaign against gene technology has no base in ecologically sound science. In the case of gene technology there is substantial evidence for positive environmental effects with decreased pesticide use and healthier food. The campaign neglects the beneficial effects of these plants for the environment. Unfortunately many environmental activists have chosen to publicise only potential adverse effects of GM crops during their campaign, natural phenomena like gene transfer or pollen movement between organisms are declared as a phenomenon related only to GM crops although this happens throughout nature". Patrick Moore, a founder member of Greenpeace, has said the present Greenpeace campaign is junk science and pagan myth.

We have in recent years been treated to flag wavers like "superweeds", "genie out of the bottle" and "frankenstein food"; statements as empty of meaning and content as those who mouth them. Superweeds are merely herbicide resistant weeds. There are over 100 weeds world-wide with resistance to some 15 different herbicides. There are even four crops with natural herbicide resistance from conventional breeding. These include oil seed rape and are used by farmers.

If you sow a rape crop with natural herbicide resistance, only marginal regulations apply and the crop could be grown alongside an organic farm without objection. The herbicide resistance genes would spread to surrounding weedy relatives by so-called gene flow although as we now know at a very low rate. Furthermore pollen from this crop could be spread by bees up to a kilometre away although it would probably not be viable at this distance. The chances of such pollen successfully competing with local sources and producing seed would be extremely remote. Perhaps more important there would be no objections from green organisations.

However in one case this natural herbicide resistance gene has been isolated and inserted back into oil seed rape by GM. Planting this GM crop necessitates satisfying 50 pages of regulations, four years of safety tests, 3–4 committees for approval with detailed examination and at the end of the day the likelihood of getting your crop trampled by unthinking activists. You would also get objections from organic farmers miles in every direction. Yet the spread of resistance genes to weedy relatives would be identical between the two crops and spraying both fields with herbicide would lead to identical ecological effects. Common sense is called for here and there is certainly a lack of common sense in current attitudes with supposed contamination by GM.

If rape is removed from the field, the herbicide resistance gene in feral weedy relatives would disappear within a few years. If the cultivated field is left fallow, both GM and non-GM rapes would disappear within a few years. Like any other crop plant, domesticated rape cannot compete with weeds. The genes we put into crops are for our benefit and not for survival in the wild. Crops last no longer than a domesticated chiuhuahua would last in the company of wolves. Populations of weeds are a sea of natural mutant variants. I am unable to think of any gene they could acquire from our efforts that would improve their weediness. Certainly in ten thousand years of plant breeding and gene flow into weedy relatives none has ever been discovered.

The "genie out of the bottle" is really attached by elastic and is easily recorked when required. When I asked the BMA for the evidence for their genie out of bottle statement made by Sir William Aesscher I was told that a lot of people were saying it!

The main goal we are told by GM opponents is to "go organic". Was this a thought-through policy or made up on the hoof? It is quite clear to me it was the latter.

Experts tell us that cancers that occur under the age of 65 are avoidable. thirty percent of these cancers are thought to result from poor diet. Over 200 detailed investigations have shown that a diet high in fruit and vegetables cuts all cancer rates by at least half. But only 10% of us eat the recommended fruit and vegetable requirements. Increasing the price of these essential foods will reduce consumption, particularly in the poorest families for which the food bill is a much higher proportion of their weekly wage. The consequence, higher avoidable cancer rates, premature death and soaring health bills. Organic food, whatever its supposed environmental merits (and incidentally these merits are shared by many conventional farms), is less efficient and more wasteful of land. For a variety of reasons it comes at a much higher price and will continue to do so. Any attempt to "go organic", to thus increase the price of fruit and vegetables and thereby reduce consumption, will have the consequences on cancer and death I have listed above. Let us hope it is not your child. My fear is that unsubstantiated claims and incorrect assumptions about organic food will lead those who strive upwards on weak incomes to buy organic but eat less fruit and vegetables because of the expense. The only justification left for buying organic food is that farmers apply less pesticide in its production. But that is precisely what the current GM crops offer us but at conventional food prices or even lower! Whose food is the real benefit now?

I am often asked what do I want to see in agriculture. Variety is probably the spice of stability. My own preference is for Integrated Crop Management (ICM), a sustainable but efficient technology organised in the UK by LEAF and CWS farming systems amongst others. ICM requires the farmer to use his intelligence whilst delivering on the so-called environmentally friendly front. In fact ICM with its emphasis on crop rotation, integrated pest management, zero tillage and precisely timed manure and mineral application is nicely placed between two extremes. The organic farmer who does what he is told by the Soil Association (something I tell students is best described as authoritarian farming), and the conventional farmer who merely does what he is told on the instruction leaflets by companies. As for many of our students with lecture information, the instructions and rules pass through, without stopping in the brains of either. The goal must be to train the farmer to view his farm as an ecosystem and then leave it to the individual and his particular circumstance to construct his own farming system. Advantageously a variety of agricultural styles would result which should improve the stability of food supply in the uncertain years ahead.

# Biotechnology's Negative Impact on World Agriculture    15

MARC LAPPÉ AND BRITT BAILEY

*Marc Lappé received a Ph.D. in pathology from the University of Pennsylvania in 1968. Following a decade of research on immunology and cancer, he joined the Hastings Center, the nation's first research institute in bioethics, where he directed its Genetics Research Group. In 1976, California governor Jerry Brown invited him to start the first ethics division in a state government, the Office of Health Law and Values. He also directed the state's Hazard Evaluation System.*

*In 1981 after being awarded a four-year fellowship from the National Institutes of Health/National Endowment for the Humanities to study human genetics and vulnerability to toxic substances, Lappé resigned from California government. From 1986 through 1992, he headed the Medical Humanities Division of the University of Illinois College of Medicine, located in Chicago. In 1993, Lappé founded CETOS (the Center for Ethics and Toxics) in Gualala, California, a nonprofit organization in northern California.*

*For more than twenty years, Marc Lappé has been a critic of the control that large corporations exert over science, genetics, and food. His writings include* Chemical Deception, Breakout: The Evolving Threat of Drug-Resistant Disease, *and* Evolutionary Medicine.

*Britt Bailey holds a master's degree in environmental policy. She is a senior associate at CETOS.*

*In their essay, Lappé and Bailey are openly skeptical of the motives of international agribusinesses in introducing GM plants around the world. They fear that such introductions will decrease biodiversity around the world, create environmental disasters, and lead to possible collapses in national economies that depend on one monoculture from GM plants.*

---

From Marc Lappé and Britt Bailey, *Against the Grain: Biotechnology and the Corporate Takeover of Your Food* (Monroe, Maine: Common Courage Press, 1998). Reprinted by permission of the publisher.

IN THE PAST, technological advances in agriculture have been made faster than the framework of regulation and public oversight have adjusted to them. We have seen this happen with over-intensive agriculture in the 1920s that culminated in the "dust bowl," and with the overreliance on pesticides like DDT and parathion for pest control. We have seen the aftermath of overuse of fertilizers, resulting in extensive contamination of aquifers and waterways with nitrates. And we have seen it with the blind export of Green Revolution technologies leading to food shortages in countries that could not afford the required irrigation and fertilizer.

We believe we are at a similar juncture with agricultural biotechnology today. Even as more and more crop releases are planned, we remain uncertain of the long-term consequences of the wholesale shift to herbicide-tolerant or *Bt* containing food crops. Researchers lack the basic motivation (although not the technology) to track where their genes go or how their chemical dependencies affect other organisms in the micro-ecosystem. We have no master plan to chart what will occur on a regional or planetary level as more and more cropland is converted to bioengineered crops. This lack of knowledge is disturbing and should give us pause.

We admit to a fundamental skepticism of the motives of the progenitors of this new technology. Many of the companies that once made environmentally unsafe chemicals like Agent Orange or PCBs have regrouped to form biological subsidies. Many have spun off new subdivisions to deal directly with biotechnology. Some like DuPont, Dow, and Monsanto have shifted into the development of genetically engineered products and formed new life sciences divisions. Typically, each company markets only the living product genetically programmed to intensify the use of its own specialized chemicals.

The fact that a chemical company has expanded into a life science biotechnology firm carries no moral weight of its own. But the word "biotechnology" has a certain euphemistic quality. "Bio," meaning life, gives the illusion that companies are producing ecologically safe, life-enhancing products. In its more literal translation, biotechnology means the application of science to life forms for commercial objectives. In essence, this anthropocentric science is necessarily neither ecologically safe nor harmful by invention, unless and until its objects of production enter the biosphere. Spread of some transgenes is virtually assured. Unfortunately, few of the seeds created by genetic engineering are non-propagative, and hence nonviable by design. Fertile transgenic pollen can and has escaped to contaminate weedy species.[1] Even this unfortunate side effect could have been offset by encoding the genes for male sterility or selfing genes that prevent fertilization. Without such built-in protections, gene flow between transgenic and native species remains a disturbing possibility.

The full panoply of effects from the release of millions of genetically engineered crop plants are presently uncertain. At one level, they may have no greater effect than do conventional non-engineered crops. At another, they may produce subtle and lasting change. Years of human ingestion of genetically engineered foods, or of agricultural ecosystems subjected to chronic exposure to chemical residues from herbicide treatments, may lead to major health or environmental changes. Certainly the ecosystems next to transgenic food crops will experience some lasting effects. But the full gamut of secondary effects remains unknown.

## Species Composition

A major concern is that the integrity of plant species will be compromised as engineered crop acreage increases and pollen mediated gene flow swamps and suppresses rare native plants that are congeners for transgenic crops. Biodiversity may also be reduced by heavy herbicide reliance. As remote as the present prospect for gene flow seems, herbicide targeted weedy species may pick up some transgenes and evolve more quickly than now anticipated. Many of these and related second order consequences of transgenic agriculture are intricately linked to ecosystem diversity.

## Biodiversity

Agricultural biotechnology threatens to decrease the number of crop plant varieties currently grown by substituting a few varieties for the many now in commerce. Take corn for example. In order to have a patent on the technology for transgenic corn, biotechnology companies have to prove that they have constructed and will maintain a *uniform product*. Such a requirement guarantees the genetic distance present among genetically engineered seeds will be kept extremely narrow. Any resulting progeny will be similar, if not identical, to one another.

Concern that genetic diversity will be sharply curtailed, even among traditional cultivars, is underscored by a series of events culminating in Pioneer Hi-Bred International Seed Company's decision not to offer transgenic seed corn. Beginning in 1995, Monsanto and Pioneer began negotiations for Pioneer to carry Monsanto's Roundup Ready™ corn. But on November 13,

1997, after two years of talks, Pioneer abruptly pulled out of the deal, stating in a letter to its customers that the contract's terms "could significantly limit the number of traits, genes, and technologies" among the types of corn Pioneer intended to market.[2] This decision marks the first time a major seed company has gone public with its concerns about the restrictive covenants and conditions of Monsanto's contracts. More to the point, Pioneer's courageous decision underscores the value seed companies place in maintaining a diverse stock of seed germ plasm, a vital necessity for ensuring crop characteristics matched to shifting growing conditions and climatological change.

Protesting Monsanto's apparently heavy handed terms, Pioneer's press release declared "we concluded that Monsanto wanted to ultimately determine what additional traits could be included in those [transgenic] products and the price to be charged to the farmer for those traits."[3] In response, Monsanto stated it was committed to making all its crops widely available and "regrets" Pioneer's news statement because it "put an inappropriate spin" on the negotiations.[4] Monsanto appears to have missed the point: it is not how many seed companies to whom they have successfully offered their technology, but how restrictive their offerings have been.

If the plants being engineered were confined to the United States, the issue of decreasing biodiversity might not be a major one. This is because the United States is not a center of biodiversity for the crops that have been commercially engineered. But most of the transgenic crops are slated for worldwide distribution in countries such as China, Thailand, India, Brazil and South Africa. These countries are in high species diversity zones, increasing the risk for biodiversity disruption. In these countries the proximity of weedy species closely related to the engineered type may permit genetic transfer of the novel engineered property from transgenic to native species.

Agricultural diversity was traditionally maintained by farmers in microecosystems sharing the best adapted crops that came from each year's planting. For thousands of years, farmers exchanged seeds allowing them to maintain a dynamic portfolio. Often the resulting broad-based gene pool proved essential to protect their fields from blights or other depredation. In many indigenous cultures farmers inherently knew the value of keeping their fields diverse. If a blight or rot attacked one variety, another would likely be immune, saving their fields from total loss. Historically, such variation and diversity have assured protection of food supplies. Why then are we moving away from this traditional goal and towards transgenic monocultures? The answer turns on economic factors.

## Commercialization

For a properly equipped farming operation, genetically altered seeds cost a premium and rely on expensive herbicides, but still save tens of dollars an acre in production costs. Transgenic seed is designed by the manufacturer to favor economies of scale, saving the large-volume farmer more money as increased acreage is devoted to engineered crops. Producers of genetically engineered crops know these facts and cater to large commercial farms. Such a tendency will exaggerate an already disturbing trend. The small noncommercial farms are being lost progressively to large commercially intensive farms on an ongoing basis. In the United States, for example, 73% of the farms reported sales of less than $50,000 and contributed only 10% of all farm income.[5] In contrast, the 2.2% of farms with annual sales of $500,000 or more accounted for almost 40% of farm income. The dominance of large scale operations leads to greater reliance on crop uniformity. And crop uniformity means reliance on fewer and fewer germ plasm lines of commodity crops like corn, wheat, or soybeans.

## Domestication of Seeds

As we continue to domesticate our seeds, the genetic diversity afforded by normally diverse genomes will be lost in the name of uniformity. This uniformity goes against the grain of evolutionary forces which have selected for plant diversity over tens of millions of years. Historical crop failures can be linked to genetic limitations (see figure).

The potato blight that caused mass destruction to the potato fields in Ireland during the middle of the nineteenth century can be attributed to diversity loss. For approximately 250 years, the potatoes in Europe descended from just two crop varieties. Potato blight, caused by a fungus known as *Phytophtora infestans*, spread throughout Europe by following the path of genetic susceptibility in these two varieties. But such losses were preventable had sufficient diversity been maintained.

The potato originated in cultivars from the Andes in South America, where hundreds of different varieties of the potato are grown. The blight that struck Europe and Ireland also took hold in the distant Andes. In Ireland 2 million people died from starvation, whereas in the Andes, only a few crops were lost. There, potato varieties survived due to the presence of genes that conferred resistance to the blight. After the epidemic, Andean wild potato relatives were widely used to restock European potato farms.

## Crop Diseases Resulting from Monoculture

| YEAR | DISEASE | CROP | COUNTRY/REGION | AMOUNT OF CROP DAMAGE | $ VALUE |
|---|---|---|---|---|---|
| 900 | Viral | Corn | Central America | ... | ... |
| 1845 | Fungal | Potatos | Ireland | 1 million died of starvation | ... |
| 1860 | Fungal | Grapes | Europe | ... | ... |
| 1865 | Fungal | Coffee | Ceylon | ... | ... |
| 1890 | Viral | Sugar Cane | Indonesia | ... | ... |
| 1916 | "red rust cut" | Wheat | U.S. | ... | ... |
| 1954 | "red rust cut" | Wheat | U.S. | 75% | ... |
| 1969 | Bacteria | Rice | Asia | ... | ... |
| 1970 | Virus | Rice | Philippines | ... | ... |
| 1970 | Southern corn-leaf blight | Corn | U.S. | 15% | 1 billion |
| 1984 | Citrus Canker | Citrus | U.S. | 18 million trees destroyed | ... |
| 1989 | aphid/insect | Wheat | U.S. | 34 million acres | 300 million |

Graphics by Platt & Company

Other epidemics among genetically uniform food crops include brown spot disease in an Indian rice crop that began the infamous Bengal famine, a wheat epidemic in 1917 in the United States, and an oat crop failure in the 1940s which eliminated eighty percent of the crop. Each time an outbreak occurred, resistant forms were obtained from the centers of diversity containing wild relatives of the crops that failed. These resistant forms were essential to reestablishing food production and insuring the survival of the species.

## Crop Uniformity

The increasing trend toward uniform farming practices that encourage growing one variety of a crop on a mass scale is almost certainly going to be exacerbated by the availability of transgenic seed. The resulting crop patterns may well increase the fragility of the crop and permit the introduction of widespread disease. This is likely because genetic diversity is integral to sustainability, balance, and the survival of crops from harvest to harvest. The genes for adaptations favoring survival evolved under conditions of diversity. Wild plants are under constant pressure from pathogens, pests, severe climates and unfavorable soils. As a result, they have evolved a myriad of strategies for survival including thorns, natural toxicity and fibrous tubers.[6] Many of these defensive characteristics, maintained as part of the reservoir of genetic diversity, are being progressively lost through domestication.

Genetic engineers drive this process still further by isolating a small subset of these traits and putting them into a selected small number of cultivars. For instance, the Liberty Link™ technology is currently in a limited number of seed stocks, greatly reducing the seed types available for sale just one or two growing seasons ago. This trend amplifies the selection pressures towards uniform traits. According to some market surveys, many consumers "want" only limited varieties for their home pantries. Such pressures further compromise the genetic variation normally maintained by natural selection. Today, protecting genetic seed banks and their associated diversity is a diminishing enterprise, maintained by only a few committed scientists who run germ line seed banks on shoestring budgets. It is worth reviewing the history of germ plasm conservation to understand the magnitude of what may be lost if we permit the genetic homogenization intrinsic to mass production of transgenic crops.

## Introduction of Pesticides

Crop uniformity carries hidden costs in heightened needs for protection. As the domination of smaller and smaller numbers of corporation selected, transgenic varieties increase, so does the risk of catastrophic disease or pestilence similar to those cited above. This dilemma, in part, encourages the proliferation of the chemical means of disease and pest control.

The pattern of increasing pesticide reliance to control insect infestations is disturbing. And chemicals do not appear to be the answer. In the last forty years, the percentage of the annual crops lost to insects and disease in the United States has doubled. Since pesticides were introduced in the 1940s, the proportion of crops lost to insects in the U.S. has grown by 13 percent.[7] By 1945, farmers were using 200 million pounds of pesticides each year in the United States.[8] Thirty years later, the annual total had risen to 1600 million pounds. Currently, global pesticide sales continue to rise. In 1996 global sales of pesticides topped $30.5 billion and are predicted to rise to $33.1 billion by the year 2001.[9]

Farmers are now not only committed to buy new seed each year, but also a set poundage of these pesticides, especially those genetically predicated on their seed purchases. With the advent of Liberty Link™ and Roundup Ready™ technologies, this pre-commitment will be even stronger. Transgenic scientists appear to be ignoring the maxim that just as insects develop immunity to pesticides, disease causing organisms also adapt to chemicals and the genetic defenses of plants. All of these organisms have coevolved by adapting to a changing environment. By losing plant varieties, plant growers are losing opportunities to breed for essential natural defenses.

## Risks of Monoculture and Monopoly

We have other concerns about this disturbing trend towards increased control over genetic diversity. When a genetically controlled monoculture of a given crop is substituted for a race of microacclimatized potatoes, corn, cotton or soybeans, this substitution imposes on the farmer a higher dependency on uniform soil conditions, higher fertilizer and water use, and machine-dependent harvesting methodologies. More critically, any blight, fungus, rot or other disease which might previously threaten a portion of a region's crops now may threaten to devastate a swath of genetically identical cultivars. The likelihood of some such apocalyptic scenario might be expected to motivate corporate concern to anticipate and monitor transgenic crop loss. Instead, an entire army of detail men have been assigned only to monitor seed use and prevent theft of transgenic varieties. This omission left an entire industry unprepared to cope with the first two major transgenic crop failures.

## Genetically Engineered Crops in Jeopardy

The first occurred when the cotton bollworm successfully overran Monsanto's Bollgard® crops in Texas during the summer of 1996. In 1997, a disturbingly similar failure afflicted a portion of the Roundup Ready™ cotton crops in the Southeast.

In the early weeks of August, 1997, farmers throughout the mid-south region of the United States watched the cotton bolls on their Roundup Ready™ cotton fall off. The resulting damage ran into the millions of dollars affecting at least 60 different farms. The failing plants, containing an inserted Roundup Ready™ gene making the cotton plants able to withstand two seasonal dousings of Roundup® herbicide, were among the first to be grown commercially. The 1997 planting season was to be the debut of this much heralded product. Approximately 600,000–800,000 acres of the newly bio-engineered crop created by Monsanto Company were sown with Roundup Ready™ cotton, or about 2.5% of the 14 million acres of cotton planted nationwide.

But three-quarters of the way through the growing season, something went awry. Cotton bolls, the billowy fruit of the plant which embraces the cotton seeds (which are ginned from the raw fibers) were lost after the second and final Roundup® application. Many of the bolls simply fell off of the plant after spraying. These apparent failings occurred in the states of Mississippi, Arkansas, Tennessee, and Louisiana, and were thought to be Roundup® related. According to Robert McCarty of the Bureau of Plant Industry in Mississippi, whom we reached at the peak of the epidemic on September 2, 1997, "we are receiving complaints from farmers every day." The complaints were all identical: the bolls became deformed and subsequently fell off the plant. Mr. Bill Robertson, a cotton specialist in Arkansas, pointed out that they experienced similar problems as well. "We call the malformation 'parrot-beaked', because the bolls look like the beaks of parrots, then they fall off of the plant before they are mature," Robertson said. The first reports of the apparent failure of the crops placed the number of affected acres in the 4–5000 range, though according to Mr. McCarty, "we are talking at least 20,000 acres in Mississippi alone, and we are getting new complaints every day. Now that is a lot of acreage, economically speaking. Some farmers are individually losing $1 million due to this problem."

As of late 1997, the investigation of this disturbing reversal of fortune remains inconclusive. At the state level, agriculture agencies are gathering data mostly for economic analysis. These agencies support the farmers, and they became involved in order for the farmers to gain compensation for their losses. Monsanto Company is also doing an investigation of economic losses, and is likely to be the only one capable of discovering why the Roundup Ready™ cotton crops apparently failed. The likelihood that they will finger their own genetic contribution to the loss is low. According to Karen Marshall in Public Affairs of Monsanto Company, "there are a number of environmental factors that can put stress on cotton plants."

Beginning in August 1997, the apparent failures did not occur in all cotton varieties in the same region, just in those few varieties that were Roundup Ready™.[10] Ms. Sunny Jeter, a Roundup Ready™ marketing representative of Monsanto, insisted that the apparent failure was only occurring in a very small portion of the Roundup Ready™ cotton crops. She emphasized Monsanto was being very proactive in getting information to farmers about the problem. Tommy McDaniel, a State of Mississippi agricultural specialist acting on the front lines, took a different tack when we interviewed him. McDaniel declared, "Monsanto is not talking to anyone and they are not saying what is causing the problem."[11]

The details that have emerged to date give little cause for optimism. Something appears to have gone wrong with the Roundup Ready™ technology. The apparent failure is occurring in specific Roundup Ready™ Paymaster (a Monsanto subsidiary) varieties #1244, #1215, #1330, and #1220. All of these varieties were used in the two previous years without any apparent problems. But in 1997, the Roundup Ready™ versions of these varieties apparently failed.[12] Several extension agents and investigators with whom we spoke speculated about why the crops are apparently failing. Most are hypothesizing that the newly inserted gene has caused instability within the Roundup Ready™ cotton crop genome, an effect made evident in the $F_3$ to $F_5$ generation of the plant.

As with the apparent failure of the genetically engineered *Bt* cotton, the 1997 Roundup Ready™ apparent failure may remain unsolved. Because of the lack of tracking and effective plant epidemiology, we may never know its causes or origins. Monsanto's view, in the fall of 1997, was simply that "the information is not available." The government does not require reporting after deregulating the crop, leaving the public and the farming community alike in the dark about the true cause of the problem.

We see a larger issue here. When Monsanto released its technology in 1997, they asserted it was "ready" for commercial scale application. But in the first year of large scale planting of two of its major crops, a significant portion of the released crops apparently failed. Should not geneticists have studied just where in the plant's genome its new gene was inserted? What occurred in the plant to make it shed its seed bolls prematurely? Should not a tracking system and "hotline" have been in place?

## Comment

These disturbing events, although limited in scope, underscore our concern that mass plantings of transgenic crops are at the least premature. If these engineered plants were any other life form, no one would have permitted their widespread introduction into the environment without an Environmental Impact Statement. Scientists still do not have the answers to fundamental questions. They do not know why certain genes "take" in their new host and others do not, or where the gene goes once it is ensconced in its new home. These questions are even more difficult to understand with crops such as cotton. Most crops manufactured today are hybrids. In most cases the first generation of seed is a known genetic entity when it is planted. Cotton is not a hybrid. The seeds that are planted often represent the fifth or sixth generation of plant descendants. With each new generation there is a reorganization of the genes. A new gene may be effective in one place within the genome but may cause another quite different reaction when reorganized the next year. In other words, Monsanto, the seed companies, and the farmers do not know for any given year where the new gene has become integrated in that year's genome, or how exactly it will affect the plant growth. Planting non-hybrid, genetically engineered plants one year after another can be very much a form of roulette.

We are thus left with disturbing questions as transgenic crops go into mass production. How much are we willing to jeopardize the evolutionary future of our food crops? How much uncertainty is generated by transgenic creation of new plants? And are we really ready to let large corporations play God in the critical area of food biotechnology?

## Notes

1. Ellstrand, N. C. 1988. "Pollen as a vehicle for the escape of engineered genes?" In: J. Hodgson & A. M. Sugden (eds.), *Planned Release of Genetically Engineered Organisms* (Trends in Biotechnology/Trends in Ecology & Evolution, Special Publication) Elsevier Publications, Cambridge. Pp. S30–S32.

2. Charles Connor, "Roundup®-tolerant corn seed ditched," *Memphis Commercial Appeal* 14 November 1997, Section I.

3. Reuters, "Pioneer will not carry Roundup Ready® corn," Des Moines, Iowa, 13 November 1997.

4. Reuters, "Monsanto seeks widespread Roundup Ready® corn use," St. Louis, Missouri, 13 November 1997.

5. Economic Research Service, USDA. *Forces Shaping US Agriculture.* July 1997.

6. Paul Raeburn, *The Last Harvest: The Genetic Gamble That Threatens to Destroy American Agriculture*, University of Nebraska Press (Lincoln), 1996, p. 96.

7. Fowler and Mooney, p. 48. The explanation why pesticide use is on the rise has much to do with the role of the pesticide. Many pesticides alleviate all pests, beneficial and harmful. Beneficial pests feed on the pests that are harmful to crops. Laws of nature have created settings whereby there are more "harmful" pests than "beneficial," otherwise beneficial insects would die off from lack of food. Pesticides generally do their job by killing everything, though their task is often short-lived. Typically, when the pesticide dissipates, harmful insects return to their food source while multiplying rapidly for their survival. Many surviving pests include those resistant to the chemical that killed their ancestors, redoubling our work. Dr. Carl Huffaker asserts that "when we kill a pest's natural enemies, we inherit their work."

8. Agrow: World Crop Protection News. 13 December, 1996.

9. Agrow: World Crop Protection News. 13 December, 1996.

10. Davis, Keith, Bureau of Plant Industry, Mississippi. Telephone interview. 16 October 1997.

11. McDaniel, T. Telephone interview. 4 September 1997.

12. McCarty, Will, Cotton Extension Specialist, Starkville, Mississippi. Telephone interview. 6 October 1997.

# The Population/Diversity Paradox                                16
## Agricultural Efficiency to Save Wilderness

ANTHONY J. TREWAVAS

*Anthony Trewavas is a professor at the Institute of Cell and Molecular Biology at the University of Edinburgh in Scotland. He earned a Ph.D. in biochemistry from University College, London, in 1964 and was a postdoctoral fellow in the School of Biological Sciences at the University of East Anglia from 1964 to 1970. He joined the University of Edinburgh in 1970, where he has taught and done research for thirty years.*

*Trewavas's lab is called the Edinburgh Molecular Signalling Group. The team that he leads there focuses on the role of calcium in signal transduction during plant growth as well as the role of calcium in all aspects of plant development.*

*Known in the United Kingdom as a defender of biotechnology and GM food, where it is difficult to champion such views, Professor Trewavas argues in this selection that increasing agricultural efficiency through use of nitrogen fertilizers and GM plants will not hurt the environment but save it. Increases in yields per acre mean less land to feed more people, thus allowing wild areas to be preserved for outdoor enthusiasts and for the benefit of future generations.*

"I KNOW OF NO TIME which is lost more thoroughly than that devoted to arguing on matters of fact with a disputant who has no facts but only very strong convictions" (Simon, 1996). The comment aptly summarizes a common experience (including my own) in dealing with technophobes. In one sense, the genetic manipulation (GM) debate can only be conducted on a level in which the participants are prepared to enlarge their knowledge and refine their views accordingly. I consequently have tried in this

---

From Anthony J. Trewavas, "The Population/Biodiversity Paradox: Agricultural Efficiency to Save Wilderness," *Plant Physiology* 125 (January 2001). Copyright © 2001 by the American Society of Plant Biologists. Reprinted by permission of the publisher.

article to provide plenty of facts that can be used in discussion with reasonable participants. My recommendation is to forget those who are not prepared to modify in any way a prepared (i.e. ideological) position.

My own view of GM is that its primary use to mankind must come initially in helping to solve fundamental problems that currently present themselves. These outstanding problems concern population, global warming, and biodiversity. In a longer version of this article I have tried also to provide some critique of current trends toward placing ecological views into agriculture in the hope of generating reasoned discussion. The full version of this article is on my web site (www.ed.ac.uk/~ebot40/main.html). All the information below can be obtained from the referenced articles, although I have not always indicated where.

## Human Population Increase

The United Nations' median population assessments are for 8 billion human beings by the year 2020 (United Nations, 1998; Pinstrup-Andersen et al., 1999); these figures are considered the most likely population scenario. The increase in the population in the next 20 years is expected to be 2 billion (35 × the population of the UK; 8 × the population of U.S.; 1.3% per year) and common humanity requires us to ensure adequate nutrition for these extra people where this is politically feasible. The largest absolute population increase is estimated to be 1.1 billion in Asia, but the highest percentage increase is expected in sub-Saharan Africa (80%). By 2020 more than 50% of the developing world's population will be living in urban areas instead of the 30% at present. Enormous problems in the production, distribution, and stability of food products will be generated (Pinstrup-Andersen et al., 1999). India is a prime example of these likely problems: 70% to 80% of the population currently farm traditionally and simply eat all that they grow. By 2025, India will be the most densely populated country in the world with 1.5 billion people and grossly swollen cities. Radical changes in Indian agriculture, transport, and food preservation would seem to be essential to avoid serious nutritional catastrophe.

An annual increase of 1.3% in food production is necessary at the present time to feed the burgeoning human population, assuming present diets remain invariant. However, richer populations eat more meat and a doubling of cereal yields may instead be necessary (Smil, 2000). Annual increases in cereal production, currently slightly below 1.3%, are predicted to continue to decline with the most serious food shortages in sub-Saharan Africa and the Middle East (Dyson, 2000). Most developing countries will have to lean heavily on imported food as they do now. Approximately 120 out of 160 countries are net importers of food grain (Goklany, 1999). In turn, a critical requirement is a genuine free trade in food, a situation that has still not been achieved.

Cropland and population are not uniformly distributed (for example, China has 7% of the world's arable land and 20%–25% of the world's population), which will exacerbate future problems. However, predicted rises in crop yields will not come about without policies that attach high priority to agricultural research (Alexandratos, 2000; Johnson, 2000), particularly as many developing countries desire self-sufficiency in food production. Worldwide funding for agricultural research has declined substantially in the last 20 years. These problems are exacerbated by diminishing cropland area due to erosion (for alternative view, see Johnson, 2000); fewer renewable resources, such as potassium and phosphate; less of, and consequently more expensive, water (by 2050, it is estimated that one-half the current worldwide rainfall on land will be used for industry and agriculture); and a reduced population working the land (Kishore and Shewmaker, 2000).

## Global Warming May Be Global Warning

We have stretched current ecosystem stability to the limit by the destruction of wilderness and fixed carbon in forests (Tilman, 2000). Continued combustion of coal and oil has ensured a steady increase in global-warming carbon dioxide levels. 1998 was the warmest year in the last millennium (Crowley, 2000). Predictions suggest average global temperatures will rise by 2°C to 3°C by 2100 with, more menacingly, increasing fluctuations in extreme weather conditions. The world climate is a complex hierarchical system and analysis and prediction lean heavily on the properties of nonlinearity, chaos, emergence, feedback, attractors, and self-organization (Stanley, 2000, and references therein). The properties of such systems are often strongly counterintuitive and at the best can only be based on probabilities of outcome (Trewavas, 1986). Simple solutions; banner waving; and using this form of agriculture and not that, which are proposed by elevating the importance of one factor without reference to the whole, are likely to produce dangerous or destabilizing results if acted on fully. Similar nonlinear difficulties attend attempts to construct world population and food production futures.

Arctic ice core analyses indicate the world climate can cross thresholds and jump to new stable temperature states in fractions of a decade (Stanley, 2000). One prediction, with a respectable probability, suggests cessation of the Gulf Stream with some worrying indications already reported in the salinity of the deep ocean (Edwards, 1999). The Gulf Stream maintains average temperatures 5° higher for parts of Europe. Cessation could be disastrous for those countries affected (including my own) and would require an agricultural revolution to be instituted in a few years. Most climate models predict a steady rise in temperature, but the accuracy of prediction is constrained by lack of detailed information. Climate change can radically alter rainfall patterns and necessitate large-scale population movement and primary changes in agriculture. Such dramatic climate changes are known to have occurred in the past in the Mediterranean region (for example, abandonment of Troy and Petra) and in parts of Meso-America in the 6th century A.D.

None of us will be immune to climate-change effects. Elevated ocean levels resulting from polar ice cap melting will ensure that substantial portions of land will disappear in low-lying areas, such as Bangladesh and Florida. Because many large cities are ports, and thus at sea level, increased floodings from weather extremes are more probable. Increased storm activity, floods, and long-term droughts (currently three years in Sahel, Ethiopia) will stretch agricultural resources and threaten local food production. Such situations may lead to wars. Two to 3 years of breakdown in monsoon patterns could, for example, cause nuclear exchange in Asia in arguments over limited food resources. Excessive heat frequently kills susceptible people and exacerbates respiratory problems. Tropical diseases such as malaria, the West Nile virus (which visited New York recently), dengue, and others may move outwards from the tropics as temperatures climb (Epstein, 2000). All this against a backdrop of variable volcanic activity known to alter climate patterns, sometimes drastically, with 100 volcanoes around the world capable of doing real damage (Crowley, 2000). Are the present fluctuations in climate the first rumblings of a breakdown in the feedback circuitry that controls global climate?

Atmospheric carbon dioxide has been increasing for over 100 years. How much of the increase of this global-warming gas is the direct result of human activities is still argued, but most have now concluded that it may be primary. Plowing up yet more wilderness, cutting down forests, or increasing the area of land under agriculture, thereby increasing the loss of fixed carbon, is no longer a viable option to solve population food problems. Furthermore, methane and nitrous oxide are far more damaging to global warming than carbon dioxide on a mole-for-mole basis. The primary land-based origin of these gases is anaerobic breakdown of organic material (particularly in rice paddies), bacterial activities in the digestive systems of cows, and microbial degradation of agricultural manure. The U.S. alone generates an estimated 1.3 billion tons of manure per year (Nagle, 1998). Some rethinking about the drive to organic farming with its heavy dependence on manure is urgently required.

The Kyoto 1997 Agreement is designed to control worldwide carbon emissions, although there is skepticism over whether such an agreement can be policed and achieved. This is not a good time for anyone to consider abandoning new agricultural technologies such as GM or to turn the clock back to organic kinds of agriculture.

## Maintenance of Biodiversity

Technological progress driven by the forces of technological change, economic growth, and trade is a prime cause of the problems facing biodiversity. The demands of an increasing human population are responsible for diversion of water, wilderness destruction, water quality problems, and accumulations of pesticide residues. Fragmentation of habitat and loss in turn places major burdens on the world's forests and terrestrial carbon stores and sinks (Goklany, 1998). Many species have been placed under stress and there is possibly a higher rate of species extinction now than previously, although this is contentious (Simon and Wildavsky, 1984). However, species extinction is not a necessary adjunct of large human populations. Relatively small numbers of human beings apparently eliminated mammoths, mastodons, the moa in New Zealand, the dodo, some 100 species (10%) of plants in Hawaii (Raven, 1993), and others some 25,000 years ago. Biodiversity has direct economic value. Pimentel et al. (1997) estimate that biodiversity contributes $100 billion to the U.S. economy each year.

## To Solve the Population/Biodiversity Paradox, It Is Necessary to Ensure Wilderness

To conserve the present ecosystems, increased food production must be limited to the cropland currently in use. Goklany and Sprague (1991) argue that conserving forests, habitats, and biodiversity by increasing the efficiency and productivity of land utilization represents a sensible alternative to sustainable development. This view is powerfully echoed by Avery (1999), who argues that recourse to less efficient forms of agriculture, for supposed environmental reasons, will result in plowing up of yet more wilderness and cutting down forest to feed the increasing population. However, the best land is almost certainly in agricultural production; what is left is usually of poor quality and likely to produce poor yields.

Smil (2000) has indicated that to feed the increase in population expected by the year 2050 with traditional agriculture (relying as it does for the basic mineral resources on limited recycling, rain, and biological nitrogen fixation) would require a 3-fold increase in land put down to crops. Tropical forests, much of the remaining temperate forests, and most remaining wilderness consequently would be eliminated with disastrous effects on atmospheric carbon dioxide. In contrast, feeding the increase in population could result in extreme damage to ecosystems unless farms are increasingly seen as small ecosystems with efficient recycling of minerals and water (Tilman, 2000). Use of renewable micro-energy sources would be beneficial. However, the Haber-Bosch process of chemical nitrogen fixation is completely sustainable if solar sources of energy are used.

Although increasing efficiency as a conscious strategy to reduce environmental impacts is virtually an article of faith for the energy and materials sector, it has received short shrift for agriculture, forestry, and other land-based human activities. Many institutions (e.g. green organizations) and strategies that would conserve species and biodiversity are conspicuously silent on the need to increase the efficiency of farmland use (Goklany, 1999). Either they do not understand the policy, or improving efficiency contradicts their desire to impose some less-efficient, supposedly ecological solution on agriculture. However, the consequence of less-efficient agriculture will be the elimination of wilderness that by any measure of biodiversity far exceeds that of any kind of farming system. It is the fundamental contradiction in current environmental arguments (Huber, 1999).

Broad technological progress is also necessary to ensure that affluence is not synonymous with environmental degradation by helping to create the technologies and financial resources needed to reduce pollution and natural resource inputs of consumption across the board. Readier availability of the necessary technology and fiscal resources will also help translate the probably universal desire for a cleaner environment into the political will for public measures.

## How Have Technological Improvements in the Past Helped to Preserve Wilderness?

From 1700 to 1993 there was an 11-fold increase in human population but only a 5.5-fold increase in cropland area. The recent improvements in agricultural efficiency brought about by technology can be seen when comparing the figures from 1961 to 1993. An approximate doubling of the world population has been gained without massive starvation and with a barely detectable increase in cropland. The agricultural yield has been a per capita increase, over and above the increase in population, and this must remain as one of the major technological achievements of the last century.

The total estimated land in use as farmland in 1993 was 4,810 Mha [1 Mha = 1,000 hectares or 2,471,000 acres—*Ed.*]. Much of this land is rough grazing and of poor soil quality with toxic levels of aluminum toxicity or low pH. But in total, 36% of the land surface (excluding polar caps) of the globe is farmed. Farming is the largest land management system on earth.

If we had frozen technology at 1961 levels, to feed the 6 billion in 2000 we would need to increase the cropland area by 80% (910 Mha), thus converting 3,550 Mha (an additional 27% of the land surface) to agricultural uses (Goklany, 1998). This calculation assumes that new lands would be as productive as present cropland, which is unlikely. The effect on atmospheric carbon dioxide levels would be disastrous. This putatively additional farmland exceeds net global loss of forest since 1961 (143 Mha) and matches the increase in cropland since 1850 (910 Mha). Ausebel (1996) estimated that wilderness the size of the Amazon basin has been saved by technological improvements since 1960. Technological improvements in U.S. agriculture in the last decades have ensured that 80 Mha of farmland has been returned to wilderness in the U.S. (Huber, 1999). If U.S. agriculture had instead been frozen at 1910 levels (part organic technology) then it would need to harvest at least an extra 495 Mha to produce present levels: more than the present cropland and forest combined.

Many technological developments have given rise to this huge improvement in yield and thus the saving of wilderness. Without pesticides, 70% of the world food crop would be lost; even with pesticide use, 42% is destroyed by insects and fungal damage (Pimentel, 1997). Dispensing with pesticides would require at least 90% more cropland to maintain present yields. Yields from irrigated fields are three times those from nonirrigated crops (Goklany, 1998). In 1960, 139 Mha were irrigated and in 1993 this amount had increased to 253 Mha. Without irrigation, 220 Mha of extra cropland would be required to feed the current population. Because application of fertilizer can increase yields by anywhere from 1.5- to 2-fold, dispensing with fertilizer would require at least an extra 400 to 600 Mha of cropland (Smil, 2000). Without these technologies, current food production would only have been achieved by plowing up an extra 2,000 Mha!

## The Downside of Technological Progress: Problems to Be Solved

Water has been diverted for irrigation and industry, but often used wastefully (Evans, 1998). On average only 45% of irrigation water reaches crops (Goklany, 1999). In 1997 the Yellow River (China) ran dry for 200 days as a result of low rainfall and extraction for industry and agriculture. The Colorado River has not reached the sea for many decades. Eutrophication and oxygen depletion caused by nitrogen and phosphate leaching from agricultural lands have resulted from the profligate use of manures and fertilizers (Smil, 1997). Stable pesticide residues are now much lower than 30 years ago because the chemical industry ensures that new pesticides are biologically unstable. Pesticide residues are detected rarely now in vegetables but it is more common that one or a few residues can be detected in about one-half of supermarket fruits at levels 100- to 1000-fold below safe recommended limits. However, current procedures for application are wasteful; only 1% of pesticides is thought to land on target.

Technological progress to solve the above problems is now necessary to help ensure that a growing human population does not squeeze out the rest of nature in the process. Abandoning technology is not the answer; improving technology to remove the hazards ensures continued benefit to both mankind and the environment. Integrated crop management systems (Chrispeels and Sadava, 1994) that optimize the use of pesticides, minerals, and water offer the best potential for future conventional agriculture to achieve yield increase without waste.

## The Environmental Transition

The vital basics of life are warmth, food security, freedom from disease, and long life. These basics require a high standard of living and people are prepared to ignore the environmental impacts of industrialization until the basics are achieved. . . . Most damaging environmental effects are associated with agriculture necessary to feed large numbers of people. No form of agriculture is really environmentally friendly because wilderness is eliminated and diversity is largely replaced by crop monocultures.

The environmental transition is marked by reductions in emissions such as sulfur dioxide (i.e. acid rain) from industry. There is also a change in perception from Mother Earth, providing an abundance of resources, to Spaceship Earth, with its limitations in provision. The "ultimate resource," human ingenuity and creativity, is not limited but increases with population numbers. The concept of Spaceship Earth is drawn from ecology and may be completely invalid for many natural resources (Simon, 1996).

Detection of environmental problems requires advanced technology and equally advanced technology and wealth to solve the problems. There will always be problems until individual ambition is satisfied. Economic growth is commonly blamed for much environmental degradation (Myers, 1997). Economic growth is not synonymous with quality of life nor an end in itself, but merely the means by which all individuals advance their quality of life for themselves and their children. But until the majority of nations pass through the environmental transition, the overall quality of the planetary environment is unlikely to improve. No government is going to agree to rules and conditions that keep their population poor. It would certainly be hypocritical for rich nations to impose constraint on others who have not yet achieved the fundamental basics of human existence. To impose such views would be tantamount to yet another example of western cultural domination. The misinformation about GM to third-world countries by current activist groups is just such an example. It is fortunate that many countries have decided to ignore the propaganda.

Living in harmony with nature, a theme of new-age groups, is a possibility that disappeared some 5,000 to 10,000 years ago and is not sought by many in poorer nations. One can, if he or she wishes, live in harmony, but one will live in poverty if one lives at all. The present wealthy and complex western societies require large numbers of people to carry out the necessary highly diverse tasks.

## Does Higher Population Growth Increase Habitat Conversion?

"The battle to feed all of humanity is over. In the 1970s and 1980s hundreds of millions of people will starve to death in spite of any crash program embarked upon now" (Ehrlich, 1968). Like Malthus before him, Ehrlich failed to appreciate that technological advances negate predictions of gloom. This time the green revolution intervened. Pressure from population increase, economic necessity, and the mere statement of the problem usually throws up solutions. It is notable that critical advances in agricultural technology, such as agricultural engineering, recognition of mineral requirements for plant growth, the Haber-Bosch process for ammonia production, and the green revolution all occurred at times in which food provision and population problems were pressing. Predictions could have been made over 100 years ago that burgeoning populations and business in London would result in the city being knee deep in horse manure (Huber, 1999). It is fortunate that the internal combustion engine intervened preventing potentially dangerous levels of ammonia toxicity!

The impact of plant breeding improvements and the green revolution (rice and wheats are responsible for much of the recent increased yield. Increased yields in India indicate the achievement. In 1950, India produced 1,635 Kcal per day per person and in 1963 produced 2,069 Kcal per day per person. The recommended minimum is 2,300 Kcal per day per person and a recommended average is 2,700 Kcal per day per person to ensure that virtually all have an adequate diet. In 1950, India produced 6 million tons of wheat and in 1998 produced 72 million tons. The total land area of India is 292 Mha and from 1961 to 1998 the population doubled to 1 billion. However, the per capita production from 1961 to 1998 actually increased from 161 to 170 Mha (Goklany, 1999).

If food production had been kept at 1951 levels (as argued by green revolution critics such as Shiva [1991]), then the requirement for cropland by 1998 would have exceeded India's land mass (thereby eliminating all wilderness and forest) or massive starvation would have been unnecessarily inflicted. In fact, Indian forest and woodland expanded by 21% between 1963 and 1999 (Goklany, 1999). The claims by Shiva (1991) that "the food supplies (in India) are today precariously perched on the narrow and alien base of the semi-dwarf wheats" have been shown to be merely polemic and have no scientific basis. The number of land races dramatically increased with the green revolution; the resistance of green revolution cereals to rust is much greater than previous varieties (Smale, 1997).

Those who constantly agitate for the worldwide introduction of primitive and frankly "land-guzzling" forms of agriculture must answer this basic question: How would their form of agriculture have fed the burgeoning human population? Although recognizing that the world produces a slight excess of food (about 8% over consumption), without the agricultural efficiency of western agriculture (the main exporters), most countries of the world would have experienced serious food shortages and the attendant human illnesses that go with starvation. Sentiment is no substitute for a full belly.

All technologies have problems because perfection is not in the human condition. The answer is to improve technology once difficulties appear; not, as some would wish, discard technology altogether. Remove the problems but retain the benefits! The benefits of modern agricultural technology are well understood; now is the time to reduce the undoubted side effects from pesticides, soil erosion, nitrogen waste, and salination. GM technology certainly offers some good solutions.

## Literature Cited

Alexandratos N (2000) World food and agriculture: outlook for the medium and longer term. Proc Nat Acad Sci USA 96: 5908–1914

Ausebel JH (1996) Can technology spare the earth? Am Sci 84: 166–178

Avery D (1999) The fallacy of the organic utopia. In J Morris, R Bate, eds, Fearing Food. Butterworth-Heinemann, Oxford, pp 3–17

Chrispeels MJ, Sadava DE (1994) Plants, Genes and Agriculture. Jones and Bartlett, London

Crowley TJ (2000) Causes of climate change over the past 1000 years. Science 289: 270–277 [Abstract/Full Text]

Dyson T (2000) World food trends and prospects to 2025. Proc Nat Acad Sci USA 96: 5929–5936

Edwards R (1999) Freezing future. New Sci 164: 6 [ISI]

Ehrlich P (1968) The Population Bomb. Ballantine Books, New York

Epstein PR (2000) Is global warming harmful to health? Sci Am 283: 36–44

Evans LT (1998) Feeding the Ten Billion. Cambridge University Press, Cambridge, UK

Goklany IM (1998) Saving habitat and conserving biodiversity on a crowded planet. Bioscience 48: 941–953 [ISI]

Goklany IM (1999) Meeting global food needs: the environmental trade offs between increasing land conversion and land productivity. In J Morris, R Bate, eds, Fearing Food. Butterworth-Heinemann, Oxford, pp 256–291

Goklany IM, Sprague MW (1991) An alternative approach to sustainable development: conserving forests, habitat and biological diversity by increasing the efficiency and productivity of land utilisation. Official Programme Analysis, U.S. Department of the Interior, Washington, DC

Huber P (1999) Hard Green: Saving the Environment from the Environmentalists. Basic Books, New York

Johnson DG (2000) The growth of demand will limit output growth for food over the next quarter century. Proc Nat Acad Sci USA 96: 5915–5920

Kishore GM, Shewmaker C (2000) Biotechnology: enhancing human nutrition in developing and developed worlds. Proc Nat Acad Sci USA 96: 5968–5972

Myers N (1997) Consumption: challenge to sustainable development. Science 276: 53–55 [Full Text]

Nagle N (1998) Fresh fruits and vegetable industry. In E Kennedy, moderator, U.S. Department of Agriculture National Conference on Food Safety Research, November 12, 1998. www.reeusda.gov/pas/programs/foodsafety/proceedings/11-12 (December 17, 1999), pp 45–47

Pimentel D (1997) Pest management in agriculture. In D Pimentel, ed, Techniques for Reducing Pesticide Use: Economic and Environmental Benefits. Wiley, Chicester, UK, pp 1–11

Pimentel D, Wilson C, McCullum C, Huang R, Dwen P, Flack J, Tran Q, Saltman T, Cliff B (1997) Economic and environmental benefits of biodiversity. Bioscience 47: 1–16

Pinstrup-Andersen P, Pandya-Lorch, Rosegrant MW (1999) The World Food Situation: Recent Developments, Emerging Issues and Long Term Prospects. International Food Policy Research Institute, Washington, DC

Raven PH (1993) Plants and People in the 21st Century: Introductory Address. XV International Botanical Congress, Tokyo

Shiva V (1991) The Violence of the Green Revolution. Zed Books, London

Simon JL (1996) The Ultimate Resource. Princeton University Press, Princeton, NJ

Simon JL, Wildavsky A (1984) On species loss, the absence of data and risks to humanity. In JL Simon, H Kahn, eds, The Resourceful Earth, pp 171–183

Smale M (1997) The green revolution and wheat genetic diversity: some unfounded assumptions. World Dev 25: 1257–1269 [ISI]

Smil V (1997) Global population and the nitrogen cycle. Sci Am 277: 76–81 [ISI]

Smil V (2000) Feeding the World: A Challenge for the 21st Century. MIT Press, Cambridge, MA

Stanley S (2000) The past climate change heats up. Proc Nat Acad Sci USA 97: 1319 [Full Text]

Tilman D (2000) Global environmental impacts of agricultural expansion: the need for sustainable and efficient practices. Proc Nat Acad Sci USA 96: 5995–6000.

Trewavas AJ (1986) Understanding the control of development and the role of growth substances. Aust J Plant Physiol 13: 447–457 [ISI]

Trewavas AJ (1998) World Population Prospects: the 1998 Revision. United Nations, New York

# A Removable Feast                                                17

C. FORD RUNGE AND BENJAMIN SENAUER

*C. Ford Runge is Distinguished McKnight University Professor of Applied Economics and Law at the University of Minnesota. Benjamin Senauer, professor of applied economics at the University of Minnesota, is currently on leave at the International Food Policy Research Institute (IFPRI).*

*In the following essay, Runge and Senauer argue that food security is necessary to grow emerging democracies with stable middle classes They assert that securing a World Trade Organization agreement that allows trade in genetically modified food is essential for the achievement of food security in developing countries, but that such an agreement must be part of a larger network of self-interested agreements among nations that increase trade for all, protect the environment, and ensure food safety.*

## Food Security and Trade

THE DEBACLE of the World Trade Organization's meeting in Seattle last year underscored how much can go wrong with world trade—and how insecure the future of trade liberalization has become. America's overreaching unilateralism offended delegations from around the world and undercut the multilateral premise of the gathering. Seattle's timing and location were equally disastrous, in contrast to the carefully planned (and relatively secluded) launch of the Uruguay Round, which began in 1986 in Punta del Este. And the industrial nations, led by the United States, did not even address one of the most vital issues: how developing countries can use technology and freer trade to better feed their populations. This need for "food security" touches on almost all the hot-button issues surrounding trade— especially agricultural trade liberalization and genetically modified (GM) food—yet the American media barely noted it.

What does food security entail? First, it involves improving a developing

---

Reprinted by permission of *Foreign Affairs* 79, no. 3 (May/June 2000). Copyright © 2000 by the Council on Foreign Relations, Inc.

nation's access to cheaper food from comparatively advantaged exporting countries. It is generally more efficient and cheaper than self-sufficiency, in which a nation tries to produce all crops that its population needs, regardless of the cost or the country's natural endowments. Food security also requires that richer countries lower their tariffs on all goods from developing countries so that emerging markets can earn cash to import the food they need. Finally, the drive for food security should tap the potential of GM technology for developing countries to both enhance nutrition and boost agricultural output.

Rather than ushering in a new era in global economic interdependence, however, Seattle exacerbated the insecurity and palpable alienation among developing countries. The influence of environmental and labor groups was hurt by the presence of their radical fringes, which confirmed the worst fears of developing countries: that turtle suits and dolphin costumes are really forms of protectionist cross-dressing. It may have been a "defining moment" for the diverse array of groups who see the WTO as a symbol of multinational corporate power, but it is difficult to understand what exactly the moment defined. The summit did nothing but highlight the disarray among policymakers over trade issues. Back when trade policy was the realm of diplomats and economic experts, at least bids and offers were made. In contrast, the Seattle battleground resembled a war of many clans—with no winners and no breakthroughs.

## Calling Malthus

Amid this impasse, a troubling problem at the heart of the trade policy debate is left unanswered. Notwithstanding current surpluses and depressed commodities prices, the world may become less able to feed itself in the 21st century. The International Food Policy Research Institute estimates that about 73 million people will join the world's population every year between 1995 and 2020, increasing it by 32 percent to 7.5 billion. Almost all the population growth will occur in developing countries, and much of it will be urban. Fortunately, per capita incomes will also increase, especially in developing countries. This will allow households to purchase more meat and animal products; demand for meat alone in the developing world is projected to double between 1995 and 2020. But to meet the needs posed by population and income growth, the world will have to produce 40 percent more grain by 2020. With yield increases slowing from the heady days of the green revolution in the 1970s, only about one-fifth of this increase is likely to come from expanding the amount of land under cultivation.

In this context, trade will be increasingly vital to food security. Because cereal production in the developing world will not keep pace with demand, net cereal imports by developing countries will need to almost double between 1995 and 2020 (to nearly 200 million tons) to fill the gap. Net imports of meat will need to increase to 6.6 million tons, or eightfold. Although many antitrade activists in Seattle advocated a return to locally produced goods, including food, the hard truth is that developing countries need freer trade to feed themselves. The United States will continue to be central to this task; in 2020, about 60 percent of world net cereal imports will still come from the United States. This role does not simply reflect American dominance as a comparatively advantaged producer of grains and livestock. Eastern Europe, the former Soviet Union, the European Union, and Australia will also substantially increase their net exports. But if any of them, notably the former Soviet Union, fails to do so, the burden of supplying the rest will fall even more on net exporters like the United States.

Assuming that production and trade keep pace with demand, per capita food availability in most developing countries will rise by about 10 percent from 1995 to 2020. Despite this, 135 million children under the age of 5 are projected to remain hungry in 2020, especially in sub-Saharan Africa and South Asia. In Africa, their number is projected to increase 30 percent by 2020. If production and trade do not keep pace with demand, a Malthusian specter of rapid population growth and dwindling food supplies could emerge. Even under the projections noted above, stagnating yield increases and growing demand will mean that real prices for food could actually rise, rather than fall as they have during the past quarter century. All these trends, which brought a boon to consumers when they lowered food prices, are unlikely to persist in the next century.

The challenge of food security is therefore a race between productivity and populations with rising incomes. Here is where trade can make a difference. It enables food—primarily grain—to move from areas of surplus to areas of deficit, allowing the deficient regions to feed themselves as long as they can pay. Expanded access to rich-country markets also increases the export earnings of developing countries by raising the cash needed to buy food and other goods. Conversely, anything that restricts this movement or reduces the ability to pay for food imports will damage this capacity.

Despite the tremendous significance of food security to trade, Seattle showed that the deep rifts over agricultural subsidies and market access, especially between the United States and the EU, remain largely unresolved. It also underscored the fact that environmentalists, who came to the negotiating table late in the Uruguay Round, now intend to be fully heard in agriculture as well as in other negotiating areas, even if they do not yet speak with a clear voice. Finally, Seattle made it clear that biotechnology—whatever it may augur for world agriculture—will be aggressively opposed as a symbol of globalization.

The protesters in Seattle ignored the fact that trade can help the much larger and more pressing issue of food production and security. The world's ability to feed itself will rely on the international community's willingness to use trade as a way of moving food from surplus to deficit regions. It will also depend on whether countries adopt policies to sustain water, land, and forests and whether farmers turn to GM crops. In short, food security will emerge either as a consensus objective of international economic policy or as another battleground among competing national interests.

## Going It Alone

A government's perception of national interest too often causes it to hoard food stocks and artificially encourage production, ostensibly to buffer consumers against food shortages and increases in market prices. Even where the international market offers a source of food at cheaper prices, dependence on external sources is anathema to many politicians and their constituents in both the North and the South. Food self-reliance, even with its demonstrably higher costs, is a popular form of nationalism. Even when countries are net exporters of food—as is the United States—it is not unusual to see protectionist regimes erected for commodities in which foreign competition is seen as a threat. Prime examples are the U.S. sugar, wool, and mohair programs, which have been defended on national-security grounds as though they were government stores of strategic metals.

The appeal to self-sufficiency is even greater where historical memories of privation and food shortage exist, as in Europe and Japan. A net food importer after World War II, Europe established a protectionist regime that encouraged domestic food production to reduce dependence on the rest of the world. Unfortunately, this system survived as Europe became a net exporter of wheat in the late 1970s. Surplus food was then subsidized for export to clear European markets. This created a domestic constituency dedicated to perpetuating both domestic and export subsidies, setting the stage for continuing battles with the United States, Canada, Australia, and others. Japan, the largest net importer of U.S. agricultural products, still clings to a rice policy that grossly subsidizes its domestic production and shuts out cheaper rice from abroad.

Among developing countries, India represents an especially striking case of the pitfalls of self-sufficiency. Efforts to raise food production and reduce reliance on imports have dominated every five-year plan since the country's independence in 1947. With substantial government subsidies to wheat and rice, largely to the exclusion of other crops, India's wheat production is now ten times what it was in 1947. Today, it is the world's second-largest producer of rice, and it ties the United States as the second-largest producer of wheat. It has reduced food imports from a high of 10.5 percent of production in 1965 to almost nothing, and it even became a net exporter of both wheat and rice for a spell in 1995. Yet behind these achievements lurk more disturbing trends. As wheat and rice production and consumption have grown, production and consumption of important protein-rich foods—chickpeas, pigeon peas, mung beans, and lentils—have fallen. In fact, from 1960 to 1995, per capita supplies of protein from all plant products increased only modestly, from 47.3 to 48.7 grams a day; supplies of critical amino-acid proteins actually fell from 9,384 to 8,790 milligrams a day. As a result, more than half of the country's population is short of energy requirements and three-quarters do not meet minimum protein requirements; 624 million Indians remain malnourished. If India truly wishes to feed its citizens properly, it must accept greater food imports as a more rational and cheaper alternative to domestic wheat and rice subsidies.

India is not alone. Self-sufficiency has reduced many nations' reliance on international trade as a source of cheaper food, allegedly on the grounds that the international market is insecure. But these schemes of hoarding and protection ultimately destabilize the international market, further reinforcing this sense of insecurity. Granted, freer global trade in food grains is not a sufficient condition for food security, especially when the low purchasing power of poor countries constrains access to these supplies. But it is a necessary step toward securing cheaper and more diverse sources of food.

Sadly, the aversion that many developing countries have to food imports has less to do with such imports' purported instability than with their governments' aversion to free markets in general. Even if countries could benefit unilaterally by opening their domestic food markets, most politicians believe that no country should unilaterally "disarm" unless other countries make matching concessions. Hence, those countries at comparative disadvantage continue to insist that their trade barriers can be lowered only when others have made concessions—despite the economic logic of comparative advantage. Each nation waits for others to make the first move toward liberalized trade before moving itself.

## Making the Rounds

Given the resistance to abandoning food self-sufficiency, countries need a mechanism for entertaining bids and offers in order to reach mutual concessions and break out of the food-security dilemma. This has been the role played by trade agreements, beginning in 1947 with the General Agreement on Tariffs and Trade (GATT), where bids and offers within and across sectors were swapped to achieve an ultimate package. Rather than realizing the neoclassical free-trade dream, GATT was about mutually managed mercantilism based on compromises. The same now applies to its successor, the World Trade Organization (WTO).

Unfortunately, agriculture has always been one of the biggest sticking points. For the first seven rounds of GATT negotiations, until the Uruguay Round of 1986–93, agriculture remained largely off the table at the behest of the Americans and the Europeans, who argued that the topic was too sensitive to be subjected to the disciplines applied to manufacturing. The agricultural export-subsidy wars of the 1980s, which were brought on by European commodity surpluses, ended this mutual silence and created the conditions needed to tackle agricultural subsidies. Throughout the Uruguay Round, European agricultural interests supported the American NGOs [nongovernmental organizations] that would do their bidding, arguing that freer trade harmed U.S. farmers as well as European ones. Yet this view was largely rejected by most (although not all) U.S. farming interests, which supported expanded U.S. agricultural exports. In the end, major liberalization still eluded negotiators, despite some modest arrangements on export subsidies, market access, and sanitary and phytosanitary measures.

At the same time, environmental issues that touched agricultural trade liberalization began to emerge. This linkage arose partly from a burgeoning perception that growth through trade would undermine environmental quality, leading to a worldwide "race to the bottom." Although this pessimistic argument was not supported by empirical evidence, it retained a large following among environmental protectionists. In contrast, more optimistic groups saw a chance to protect environmental resources by using liberalization to enforce environmental oversight. But neither side dealt explicitly with how trade affected agriculture or food security—until the emergence of the GM debate.

GM food has now become a cause célèbre for consumer groups that had been relatively uninvolved in trade policy, lending strength to a coalition of antitrade activists that includes labor, environmentalists, and left- and right-wing groups anxious to protect national sovereignty. This new combination of forces has successfully used the "Frankenfood" issue to mask its protectionist elements while posing as the enemy of corporate multinationals.

The initial opposition to GM food grew in Europe. It aimed first at American multinationals like Monsanto, one of the leading sellers of genetically modified corn, soybeans, and cottonseeds. When Monsanto indicated the possibility of a "terminator" gene that would render the offspring of GM plants sterile (thereby preventing farmers from producing seeds for replanting), a coalition of developing countries and European farmers formed to develop new trade barriers against the spread of these crops. Environmental pessimists now claim that GM technology could spread unwanted resistance to weeds and insects beyond the target species, potentially creating "superweeds" and other unwanted ecological side effects. For their part, consumer groups argue in fairly vague terms that GM food poses a threat to basic health and food-safety regulations. These activists also see GM issues as useful in getting traction from other groups to oppose trade liberalization, whether or not these allies are concerned with food.

The new opposition threatens much more than the bottom line of companies such as Monsanto. It creates new problems for exporters of GM crops in the United States and elsewhere who had adopted the technology with enthusiasm, and for researchers who bet millions of dollars on the potential to help address disease and production issues for developing-world farmers. Most alarming is that these groups fail to understand what this discovery could do for developing countries. For example, new technology that raises beta carotene levels in rice—the world's most widely consumed grain—could effectively wipe out Vitamin A deficiencies within a decade. This would affect the lives, and prevent the deaths, of millions of poor children in developing countries—if the technology can be successfully transferred to traditional rice growers.

Despite the questions surrounding it, the rapid adoption of GM technology since 1996 suggests that it greatly appeals to farmers, at least in developed countries. Since then, the United States has seen the rapid commercial introduction of GM corn, cotton, tomatoes, and soybeans. By 1998, more than 500 genetically modified plant varieties were available in the United States, accounting for 28 percent of the land (2.57 million hectares) devoted to maize, soybeans, and cotton. Other countries, led by Argentina and Canada, also began planting hundreds of thousands of hectares with GM crops. These crops rapidly entered the supply chain for processed foods using corn, soybean, or cottonseed oils. Today, some 70–100 percent of processed foods everywhere may contain some GM material.

In the developing world, the appeal of GM food remains uncertain despite the long-standing efforts of major funders, such as the Rockefeller Foundation, to harness the technology to aid poor farmers. Unfortunately, most developing countries possess few technical resources to develop their own scientific and management capacity for biotechnology. Doing so would need substantial flows of capital, human resources, and scientific information and expertise across national borders. Multinational companies heavily invested in GM technology could help by establishing training fellowships for scientists, bolstered by international agreements to protect the intellectual property rights of both companies and developing countries. Yet even if the South clears the technological hurdles to developing GM food, it may face continuing NGO opposition in the North.

The GM issue connects agriculture, trade, the environment, and food security to form a complex relationship that cries out for a global structure of rules and disciplines. This is precisely what the much-maligned WTO system can provide. At the same time, these "Frankenfoods" have become central to the new protectionist case against the world trading system. The only way out of this quagmire is for the WTO to incorporate the successful concession-based approach of the past and tie food security and GM issues into a broader framework of regulations for trade, intellectual property, and the environment. The global problem posed by food security is inextricably linked to the development of the rules and agreements that operate at a level higher than the nation-state. Food security is a problem of collective national action that can be pursued only through multilateral policies, just like international commerce or environmental issues.

## Give a Little

To provide all these global collective goods, nation-states must be willing to grant concessions through negotiated agreements. In turn, these concessions should be seen as reciprocal contributions to a balanced package. For this tactic to work, countries must develop new forums or build up existing institutions to maintain the necessary multinational infrastructure.

The first and most obvious step is to secure the commercial concessions under the terms of the WTO's next round of multilateral negotiations. In particular, progress must be made in agriculture toward increased market access and reduced export subsidies. Like the Uruguay Round, the next set of trade negotiations will face major resistance from farming interests, especially in the EU. Nevertheless, increasing food security will required agricultural trade liberalization—and over time an end to the price instability generated by tariff distortions.

Although recent measures such as Congress' Africa trade bill promise additional assistance to the poorest countries, the effectiveness of development aid is dwarfed by the potential for significant increases in access by all developing countries to rich countries' markets. This will require the dismantling of protectionist regimes in the United States and the EU for sugar, peanuts, textiles, and other commodities in which many developing countries hold comparative advantages. If these countries are allowed to expand their exports—rather than receive handouts—they will find a new engine of growth. A 10 percent increase in market access to U.S. sugar markets for Caribbean producers would do more to raise incomes in the Caribbean basin than has all of the development assistance provided in the last 25 years.

Second, NGOs are correct in pointing out that new multilateral institutions must learn to deal with environmental challenges. They are wrong, however, in believing that the WTO should bear the blame. It is unreasonable and unwise to expect the WTO to assume responsibility for environmental issues unless they impinge directly on trade. Even then, it is doubtful that the WTO can tackle the manifold complexity of international environmental issues. For these reasons, many environmentalists have now joined the former director general of the WTO in calling for a separate entity to address the need for rules on ecological interdependence, just like the WTO addresses the need for rules on commercial interdependence. A new "Global Environmental Organization," for example, would create a central authority to organize the hundreds of existing environmental agreements and protocols.

Among other things, such an organization could assess the environmental implications of the expanding market for GM foods. Governments and the private sector will likely need to respond to calls for the labeling of foods and seeds. A system is therefore needed that could effectively use and develop GM technologies while allowing consumers to reject them if they wish. A Canadian survey of 8 countries found significant variation in consumer attitudes. For example, although 68 percent of all respondents said they would be less likely to buy groceries labeled as GM products, national responses ranged from a low of 57 percent in the United States to 82 percent in Germany. But the combination of consumer choice with freer trade would remove the chance that the GM issue could be exploited for protectionist purposes; consumers could choose between GM food and organic products without resorting to trade discrimination. As Alexander Haslberger, a leading European expert on biotechnology, noted in a recent contribution to *Science*, the significant public opposition to GM food will require that the industry adopt honest and appropriate labeling if it wants to avoid consumer resistance. One possible multilateral response could be under the auspices of the new biosafety protocol—or the U.N. Food and Agriculture Organization's Codex Alimentarius—to harmonize differing national standards.

The Montreal talks last January, when more than 130 countries agreed on the Biosafety Protocol to the Convention on Biological Diversity, were a good start. The protocol discusses the environmental risks and benefits in biotechnology and creates a framework to protect biodiversity in developing countries. But many unanswered questions remain. Most prominent is whether the new protocol will allow a protectionist loophole for a "precautionary principle" that bars GM-food trade even if scientific evidence of harm is insufficient. Another central issue is the balance between trade restrictions justified on environmental or health grounds and the larger obligations of nations to trade without discrimination under the WTO.

Last, the issue of food security itself cannot be used as an excuse to restrict market access. Nor can it be used to subsidize production in ways costly to the countries that need trade liberalization the most. But to reassure the countries and their citizens who are fearful of market forces, rules must be in place to provide guaranteed access to food in times of emergency. This can be accomplished by multilateral grain-sharing agreements that guarantee emergency concessionary terms. Here, as elsewhere, governments and the private sector must make a collective commitment to allay the fears of developing countries and address the calls of those most mistrustful of market forces and the dark side of globalization.

Since developing countries now account for three-fourths of the WTO member nations, new trade agreements will not be reached without their support. The developing world has the power to block future WTO accords that they perceive as hostile to their interests. Given that food security is a major concern in many of these countries, a trade commitment to enhance this basic need could generate the goodwill necessary among developing countries to facilitate their cooperation across a range of global issues. A precondition for successful international cooperation is that all participants perceive a net benefit. True, not all countries gain equally from every international accord; in some instances, they may lose on specific issues. But commitments to food security could provide enough gains seen as necessary by developing countries to win their cooperation on a range of other issues important to the industrial nations, such as the environment and intellectual property rights.

Henry Kissinger remarked after Seattle that President Clinton "could have used the occasion to put forward a farsighted program for dealing with what portends to be one of the greatest challenges of the new century: the huge gap between the sophistication of . . . globalization, and traditional political thinking still based on the nation state." Bridging this gap has thus far escaped the presidential candidates, whose international views are simply extensions of domestic interest-group politics. A larger and more comprehensive multilateral vision, which recognizes a legitimate and growing role for developing countries—and food security in particular—would ultimately benefit the United States. Realizing this vision will require more and better international institutions, not fewer and worse ones.

# From Global to Local
## Sowing the Seeds of Community

18

HELEN NORBERG-HODGE, PETER GOERING, AND JOHN PAGE

*Helen Norberg-Hodge directs the International Society for Ecology and Culture (ISEC). She was educated in Germany, Austria, England, France, and the United States, as well as her native Sweden. Her formal training was in linguistics, including work at the Massachusetts Institute of Technology with Noam Chomsky. Norberg-Hodge has lectured extensively on environmental issues in both Europe and North America, and she wrote* Ancient Futures: Learning from Ladakh, *which explores the root causes behind today's environmental and social malaise. In 1986, she received the Right Livelihood Award, commonly known as the "Alternative Nobel Prize."*

*Peter Goering received his M.A. from the Energy and Resources Group of the University of California at Berkeley, a department that conducts interdisciplinary research into the relationship among resources, society, and the environment. He is currently preparing his doctoral dissertation, titled "Sustainable Development and the Contradictions of Modernism," while also working as ISEC's research coordinator.*

*John Page trained as a barrister in London. For the last decade he has coordinated the technical, educational, and agricultural activities of the Ladakh Project, and is now programs director of ISEC. He is the producer/director of two recent films:* The Future of Progress, *a compilation of interviews with internationally known environmentalists, and* Ancient Futures: Lessons from Little Tibet, *a documentary based on Norberg-Hodge's book* Ancient Futures: Learning from Ladakh.

*The following essay is taken from Norberg-Hodge, Goering, and Page's book,* From the Ground Up: Rethinking Industrial Agriculture.

---

From Helen Norberg-Hodge, Peter Goering, and John Page, *From the Ground Up: Rethinking Industrial Agriculture* (London: Zed Books, 2001). Reprinted by permission of the publisher.

## Seattle—a Watershed?

FLYING OUT TO JOIN the non-governmental organisations protesting in Seattle in November 1999, I was amazed to find that *The Economist* had devoted its cover story to the growing popular resistance to the World Trade Organisation (WTO) Millennium Round negotiations and to globalisation in general. We critics, the magazine claimed, were 'winning the battle for public opinion', with 'protectionist sentiments growing rapidly around the world.' As committed proponents of the 'free' trade dogma, *The Economist* was clearly alarmed that public dissent might increase the likelihood of a political stalemate that would prevent further deregulation and so 'fatally undermine the WTO'.[1] For campaigners used to working on shoestring budgets against the overwhelming might of corporate and government interests, this was heady stuff indeed.

As a founder of the International Forum on Globalisation (IFG), I was due to speak at an IFG conference held in Seattle's symphony hall, with seating for 2,500 people. We had worried that we might not attract anything like that number. At the event, not only was the hall packed—by a wildly enthusiastic audience applauding, it seemed, every word from speakers such as Vandana Shiva, Martin Khor, myself, and many others—but some 2,000 people had to be turned away. By all accounts, what Americans call ticket 'scalping' was rife outside, with $10 tickets selling for $40. This was no rock concert nor heavyweight prize-fight; this was a teach-in on trade! As the Chairman of the US Federal Reserve Bank warned the global economic community: 'Beware the rumbling out there!'

But what does this rumbling signify? According to some politicians and commentators, it is merely the sound of a confused public stumbling through arguments about free trade and the global economy, and reacting 'emotionally' against endeavours such as the genetic modification of food. This sort of thing should be left to the 'experts', they insist; it is just too complex for 'ordinary people' to understand.

Complex it is; nevertheless, the great majority of protesters knew perfectly well why they were so incensed. Their criticisms were generally articulate and intelligent. When I was asked to speak on a Seattle radio show, a farmer—the head of a rural crisis centre in Kansas—turned to me and said authoritatively, 'You know, it's the free trade treaties that are killing us.' People around the world are rapidly waking up to the fact that the promise of jobs and universal prosperity through global trade is a hollow one.

Time and again, people had told me that it would be impossible to get anyone interested in big, complicated issues like trade policy, or in long words like 'globalisation'. Well, now it's happening, as part of an extraordinary alliance of environmentalists, human rights activists and people concerned with economic insecurity and the loss of their jobs, in both North and South. This has never happened before to anything like this extent, and I believe it is irresistible. The feeling at Seattle, above all else, was that resistance to the corporate-led global economy is a process that cannot be stopped.

One of the major pathways to a more critical stance on globalisation—feeding the anger that drove protesters onto the streets of Seattle—has been the increasing concern about the genetic manipulation of food. Much of the enthusiasm seen in Seattle was engendered by the strong stance that consumers have taken against genetically modified organisms (GMOs), particularly in Europe, and by the victories that they won in the previous couple of years. In the original edition of this book, we warned of the coming marriage of biotechnology and unregulated global trade. This cornerstone of globalisation is attracting ever more attention, as movements opposed to both 'Frankenfoods' and 'free' trade continue to grow, and as people become aware of the intimate links between the two. Around the world, an increasing number of people see this unholy marriage as constituting a severe threat to food security, the environment and human health. Both biotechnology and 'free' trade are integral components of the global industrial food system, which is not only leading to a fundamental reorganisation of how our food is produced and marketed, but is also creating lifeless rural areas and overcrowded, energy-intensive cities, while at the same time severing our direct relationship with the natural world.

## The Global Industrial Food System

Over several hundred years, thousands of diverse, locally adapted agricultural systems around the world have been replaced by a single, industrial food system. Among the indicators of this shift are a dramatic reduction in the number of farmers and a concomitant expansion in the size of farms, a huge increase in the size and scope of agricultural markets, a startling decrease in agricultural biodiversity and a tightening control by transnational corporations over the world's food supply.

The negative effects of industrial agriculture are clearly spelled out in the pages of this book. For several decades now, there has been much research into the effects of pesticide residues on soil and agricultural ecosystems, and upon human health. Increased awareness of these problems has helped to spawn a fast-growing demand for organic foods, and has stimulated research into alternative agricultural methods, including traditional systems of agriculture. Much less articulated, however, are the devastating effects of lengthening the distance between producer and consumer—effects that are not addressed simply by avoiding the use of pesticides and other chemicals.

Foremost among these effects is a huge increase in the transport of food, and its attendant ecological costs. It is no longer in doubt that greenhouse gases are altering the global climate—making weather everywhere not simply warmer, but more unstable, unpredictable and extreme. The stakes are so high that to continue 'business as usual' seems irrational at best, especially when it means encouraging people everywhere to consume food transported thousands of miles, instead of food produced next door. It is a notion that borders on lunacy. Yet this is exactly what government policies in almost every country promote, and as a consequence, food miles within the industrial food system have risen astronomically in the last half-century. A typical plate of food in the United States today has accumulated some 1,500 miles from source to table.[2] On average, each item of food now travels 50 per cent further than it did in 1979.[3]

There is little merit to the argument that all this food transport simply enables people to consume fruit, vegetables and other foods unavailable from nearby sources. In 1996, for instance, Britain exported 47 million kilogrammes of butter, while *importing* an almost identical amount, 49 million kg. The situation is almost as bad for milk: of the 173 million litres of milk Britain imported, a large portion was unnecessary, since 111 million litres were also exported.[4] Figures are similar for other commodities, and for other countries. For the most part, this excessive transport benefits only a few large-scale agribusinesses and speculators, which take advantage of government subsidies, exchange-rate swings and price differentials to shift foods from country to country in search of the highest profits. Although proponents of 'free' trade argue that fleets of cargo ships, trucks and planes carrying the same commodities in opposite directions somehow lead to economic efficiency, the current system is, by any reasonable measure, absurdly *inefficient*. As economist Herman Daly has pointed out: 'Americans import Danish sugar cookies, and Danes import American sugar cookies. Exchanging recipes would surely be more efficient.'[5]

Foods grown locally have other energy-related advantages over industrial foods produced hundreds or thousands of kilometres away. Since local foods are more often consumed fresh, they usually require less packaging, processing and refrigeration: a frozen carton of peas, for example, requires 2.5 times as much energy as fresh peas, while an aluminium can of peas uses 4.5 times as much.[6] Furthermore, there is the problem of what to do with the waste that results from all the packaging required by industrial foods. In the UK, at least a quarter of household waste is packaging, two-thirds of which is from food.[7] More and more land must be devoted to burying this huge amount of waste, because it is produced on a scale that natural processes cannot possibly absorb. Much of the packaging is non-biodegradable plastic, and even paper cannot break down in dense, poorly aerated landfills.[8] Burning all this refuse is an even worse option: trash incinerators contaminate the air with hundreds of pollutants, including carcinogenic substances such as dioxin, while leaving behind an ash contaminated with heavy metals and other toxins.[9]

Growing for faraway markets is also eroding the nutritional quality of our food. Fruit and vegetable varieties are bred with characteristics to suit the global marketing system, and nutritional content is not one of them. For industrial foods, hardiness under monocultural growing conditions and the ability to transport and handle well are valued more highly than nutritional content. And because the vitamins in almost any food are gradually lost from the time of harvest, even 'fresh' foods from the industrial system are less nutritious if they have been harvested days or even weeks before reaching the kitchen table. Tomatoes, for example, are often picked green and hard so that they can survive mechanical harvesting and long-distance transport, and then ripened in rooms pumped full of ethylene gas, which artificially initiates ripening. These tomatoes are much less tasty and nutritious than the ripe tomato from the local farm or back garden, plucked from the vine and eaten the same day.

Another high priority is visual 'perfection'. Decades of agribusiness and supermarket advertising—combined with numerous senseless regulations—have persuaded people that fruits and vegetables must conform to narrow standards of size, shape and colour. Customers expect to find only bright red, unmarred apples, potatoes that are 'properly' shaped and without blemish, and carrots that are adequately large, straight and orange. Most western consumers are now so disconnected from agricultural reality that heirloom varieties of unusual shape or colour are widely unrecognised and not considered to be real foods at all. And food grown in living soil where insects are allowed to survive—sometimes leaving their mark on the produce—is considered substandard, even though it is likely to be better tasting and more nutritious than its more 'perfect-looking' industrial cousin.

The global industrial food system is as damaging for farmers as it is for consumers. One of the most conspicuous features of the system is the shrinking percentage of the price of food that farmers receive. In part, this is because a large number of corporate intermediaries—international traders, food processors, distributors and supermarkets—are taking an ever-bigger cut. In the United Sates, the real price of a market basket of food has increased about 3 per cent since 1984, while the farm value of that food has *fallen* by more than 35 per cent.[10] Today, only 21 cents of every dollar spent on domestically produced food in the US goes to farmers, the remaining 79 cents going to middlemen and marketers.[11] With that 21 cents, farmers have to pay for inputs, farm labour, machinery, rent, and so on. Many are lucky to break even after a season of hard work.

Farmers are being economically squeezed in other ways as well. 'Free' trade policies force farmers to compete with others around the world, often in places with more favourable labour costs or climate. What is more, the market for an export-oriented farmer's production can suddenly evaporate owing to currency fluctuations or a recession thousands of miles away. In the Third World, farmers are being persuaded to commodify their production, growing for the market rather than for themselves and their own communities. This ties them to the same unstable global market forces. The pressure to commodify leads to monocultural production, which also leads to greater instability. As part of this process, family members (usually men) are pressured to sell their labour in the money economy, and frequently leave the farm to seek work in the city. The consequent loss of labour on the farm creates problems for the family members (usually women and children) who remain behind. Eventually, many farming families are forced off the land altogether, leaving their ancestral homes and rural communities for the anonymity of the big city. Here they will almost certainly exist on the economic margins, living in one of the Third World's ever-expanding urban slums.

In both the Third World and the West, much industrial agricultural work is done by hired labourers, often immigrants, who have very little power or control over their working conditions. They typically perform monotonous labour, for poor wages, and are often exposed to harmful chemicals, being forced to handle produce that is covered with pesticide residues. Their living conditions are often abysmal. If they are not given (typically substandard) housing by their employer, they must seek out similar housing in town. In addition to these harsh living and working conditions, they are often treated with little respect by other social groups. Rural peoples are often seen as 'backward' and stupid. If they are immigrants, they may face even harsher persecution. Economic hardship can bring a backlash against these 'outsiders', who have left their own countries only because economic circumstances forced them to.

The non-farming members of rural communities lose out as well. As farms consolidate and rural people leave for the city, the local businesses that supported the farm community see their customer base shrink. As small town high-streets become more and more drained of life, large chain retailers, such as Wal-Mart, set up shop to serve the remaining population. These large retailers serve a much wider catchment area, making it necessary for rural people to travel further to satisfy their basic shopping needs, and contributing significantly to the pressures on small, local businesses. Evidence of this trend has been the rapid decline of specialist grocers, butchers, bakers and fishmongers, with a total loss of some 6,000 shops in the UK between 1990 and 1995.[12] Family-run grocery shops now account for only 12 per cent of the UK vegetable market, whilst the supermarket share has grown from 8 per cent in 1969 to 72 per cent in 1995.[13]

## Destroying Diversity

The consequences of the global food system for the real diversity of available foods have been brutal and rapid. Each *quartier* of Paris in the 1970s had its own colourful market, selling wonderful fruits, all kinds of vegetables, meats, superb cheeses and wine. This diversity originated at no great distance, most of it coming from different regions of France, if not from the immediate surroundings of Paris. Today it can be difficult to find garlic in Paris that has not travelled from China. In the supermarkets, Chilean grapes and Californian wines are increasingly commonplace. The diversity of French foods is in decline, and they are becoming ever more expensive.

Similarly, in the villages of southern Andalusia in the 1980s, almost all the food sold came from the villages themselves or the immediate region: goat's cheese, olives and olive oil, grapes, fresh and dried figs, wine and many different kinds of meat. Today you will find hardly anything on sale that originated in the village. Olives may come from the surrounding region, but they will have been sent to the metropolis to be packaged in plastic and then returned. Virtually everything sold is vacuum-sealed in layers of plastic. Even cheese rinds are now made of plastic.

It might be argued that, the issue of unnecessary transport aside, an increase in trade invariably leads to increased choice for consumers. After all, are not consumers now exposed to all kinds of new and exotic foods in their supermarket, foods that cannot be produced locally? While in a sense this is true, the reality is that the purported diversity offered by the global economy and its supermarkets is based on modes of production that are condemning producers to monoculture. The perceived expansion of choice obscures a startling loss in real diversity. Whereas a few decades ago, one could find more than a dozen varieties of apple or tomato in the local market, today there are typically just three or four. These are the same three or four varieties that you would find in a supermarket on the other side of the country, or on another continent—the few select varieties that have the characteristics that are most conducive to the demands of long-distance transport. As part of this process the diverse cheeses from France, the apple varieties of Devon and the olive groves of Andalusia are progressively being replaced by standardised hybrids, to suit the long distance, large-scale marketplace. Examples of the once-famed range of more than 200 regional English apples are now a rarity in English shopping baskets—an example of what 'choice' actually means in the global economy.

This lack of real diversity at the market is mirrored in an increasing number of the world's agricultural fields. The consequence of more and more farmers orienting their production towards far-away markets has been a startling loss in agricultural biodiversity. In China, of the 10,000 wheat varieties in use in 1949, only 1,000 remained in the 1970s. In the US, 95 per cent of the cabbage, 91 per cent of the field maize, 94 per cent of the pea, and 81 per cent of the tomato varieties have been lost. According to the UN Food and Agriculture Organisation (FAO), approximately 75 per cent of the world's agricultural diversity has been lost in the last century.[14] The implications of this trend for food security are ominous. Not only are there fewer kinds of foods being raised and eaten around the world, but diversity within the few remaining staples is being lost as well. Since a field planted in monoculture is more susceptible to devastation by pests and blight, the risks rise exponentially when much of the entire planet's arable land is planted in virtually identical strains. In 1970, for example, 80 per cent of the corn planted in the US shared a common genetic heritage. When a maize blight struck, it quickly destroyed more than 10 million acres of corn.[15]

## Globalisation and the Dogma of 'Free' Trade

The global industrial food system arose within the context of the corporate-led global economy. Employing taxpayers' money, and at the behest of big business, governments have funded the motorways, high-speed rail links, tunnels, bridges and communications satellites that make it possible to sell foods from the other side of the world at a lower price than local produce. Taxpayers also subsidise the aviation fuel and energy production on which this long-distance trade depends. The continual expansion of these infrastructures, along with the increased global trade that results, leads to the breakdown of local and regional economies, as they become tied to one global system. Since the 1940s, international trade has come to dominate a growing number of national economies. In this period, world trade has grown twelve-fold. Imports and exports now make up a much larger proportion of economic activity than ever before, with international trade amounting to some US$5.5 trillion annually.[16] This explosive increase in global trade has fed the growth of the trading bodies—transnational corporations (TNCs)—many of which have more economic clout than entire nations. In fact, a comparison of national GDPs (Gross Domestic Products) with the annual revenues of TNCs shows that half of the 100 largest economies in the world are now corporations.[17]

The outline of today's globalised economy was established at the 1944 Bretton Woods conference, where western leaders met to design a new financial architecture for the post-war period, in order to keep the growth economy alive and prevent another depression. This was to be accomplished by drawing more of the world's economies into the orbit of the consumer economy, thereby dramatically expanding the market for industrial goods while assuring unfettered access to the planet's natural resources. With this end in mind, three supranational institutions were established: the World Bank, the International Monetary Fund (IMF) and the General Agreement on Tariffs and Trade (GATT).

In the Bretton Woods scheme, the World Bank would provide funding for major 'development' projects, including huge centralised energy plants, long-distance transport networks, and high-speed communications systems, thereby creating the physical infrastructure required by huge trading corporations. The IMF would work to impose a standardised economic architecture—with Western-style consumer growth as the foundation—on every national economy. Strict 'structural adjustment' policies would be imposed on borrowing countries that did not adhere to that plan. GATT, meanwhile, would serve to increase every nation's dependence on long-distance trade by keeping tariffs low, removing other perceived 'barriers to trade', and turning ever more realms of life into globally tradeable commodities.

For the South, this framework ushered in the era of 'development', with goals and policies that followed seamlessly from the colonial era that preceded it. Though ostensibly more noble in its aims, the results of this framework were strikingly similar: a Northern economic model, based on industrial production, trade and economic growth, was to be systematically imposed throughout the Third World. As in the colonial period, a large portion of the South's production and resources would continue to flow Northward.

In 1994, member nations of GATT created a new and powerful governing body, the World Trade Organisation (WTO), to set trade rules and settle disputes. Member countries which join the WTO implicitly agree to reorganise their national economies in ways that are more conducive to foreign trade and investment.[18] This reorganisation includes the privatisation of industry and the dismantling of any social programme, or labour, environmental or health regulation that could be interpreted as a 'non-tariff' barrier to trade. In this context, practically any law or regulation that a government may have adopted in order to protect its environment or its citizens—such as laws regulating the harvesting of tuna to ensure that dolphins are not killed in the process, or laws forbidding the sale of manufactured goods that were produced using child labour—may be perceived as GATT-illegal, with the offending country forced either to repeal the offensive law or to pay huge fines to the country that brings the complaint.

At Seattle in 1999, trade ministers were meeting to negotiate rules for a new 'Millennium Round', which would further eliminate hindrances to the speculative flow of capital and trade around the globe, and would place greater limitations on any government's ability to protect its citizens and environment.

Part of the Seattle agenda was to develop further the 'Agreement on Agriculture', bringing the production of food more firmly into the global trade arena—thereby under the control of TNCs—and forcing governments to remove tariffs and subsidies designed to support domestic producers and maintain the integrity of their countries' rural areas. Also on the current trade agenda are Patents on Life and Trade-Related Aspects of Intellectual Property (TRIPs). Under current rules, biotech TNCs can essentially develop and patent marketable products in the lab, using genetic material and centuries-old traditional knowledge from anywhere on the planet, without giving compensation to the people or communities from which the knowledge or genetic material was obtained.

Both of these issues are of crucial importance to Third World countries, in which a majority of the population still lives on the land, and where most of this valuable genetic material and traditional knowledge resides. For Third World governments, the first priority is food: they assert the right of every country to guarantee the production of food for its own people and the right to support its own farmers. They insist that countries should have the democratic right to establish their own national economic development policies, including agriculture policy. Most are also strongly opposed to the patenting of life forms, and support the conservation of their own countries' plant genetic resources. Within the current WTO framework for negotiations, the most economically powerful countries exert the most influence; most Third World delegates are essentially bullied into signing agreements against their countries' interests—in which they had no hand in writing. In Seattle, however, these delegates demanded to be heard. Emboldened by 'the rumbling out there', they jointly declared that they would not agree to any new agenda for a new round of WTO talks. Collapse of the talks thereby became a certainty.

The current orthodoxy is that the liberalisation of trade, through the reduction of tariff and non-tariff barriers, increases competition and efficiency, releases more productive resources for growth, and generally leads to a higher 'standard of living' for all people. But all over the world, in both North and South, this trend towards 'liberalisation' is eroding laws that are designed to protect citizens and the environment, vastly enhancing the ability of TNCs to roam the planet freely in search of places where labour, environmental and health standards are weakest, and where they can extract the most surplus for the lowest cost. Governments, whose economies are increasingly linked to unstable market forces, have become beholden to these businesses, providing them with huge financial incentives to set up shop in their neighbourhood, and developing the infrastructures necessary for global trade (all at taxpayers' expense). In this process, governments have become increasingly less answerable to the needs and desires of their citizenry. Furthermore, economic 'liberalisation', along with huge government expenditures on trade infrastructures and other subsidies for big business, has created an economic playing field on which small-scale, local producers cannot possibly compete. This has led to the loss of livelihood for millions upon millions of small farmers, fishermen, producers and shop-owners. Rising GDP, large corporate profits and booming stock markets obscure the declining quality of life for a majority of the world's people.

## Genetically Engineered Food

A major new tool in the service of the global industrial food system is biotechnology. GMOs came to widespread public attention for the first time in 1996, when the media began to carry stories of how soybeans grown in the US were genetically engineered by Monsanto to be resistant to their best-selling herbicide 'Round-up'. Over 40 per cent of the soybean harvest is exported, often mixed with conventional soybeans. The American Soybean Association rejected calls to keep the genetically modified soya separate on the basis that it was 'substantially equivalent' to ordinary soya.[19] The theory of 'substantial equivalence' has been at the heart of international guidelines and testing of GMOs. According to this principle, selected chemical characteristics are compared between a GMO and any non-modified variety within the same species. If the two are closely similar, the GMO does not need to be rigorously tested, on the assumption that it is no more dangerous than the non-modified equivalent. (Conveniently for biotech corporations, while a GMO is deemed to be 'substantially equivalent' to its non-modified cousin when it comes to testing and regulation, it is somehow deemed to be substantially *different* when it comes to patenting rights.)

Unfortunately, 'substantial equivalence' is flawed as a basis for risk assessment. GMOs may contain unanticipated new molecules that could be toxic. In 1989, 37 people died in the United States after consuming a food supplement called L-tryptophan that had been produced from genetically engineered bacteria. It was regarded as 'substantially equivalent' and passed as safe for human consumption.[20] A variety of enzymes produced from genetically engineered micro-organisms is used throughout the food industry. None of these foods has been subject to the kind of long-term safety studies applied to new drugs.

The idea of 'substantial equivalence' grows out of the argument that, because people have been inter-breeding plants and animals for many thousands of years, genetic engineering is simply an extension of traditional practices, rather than a radical new method that requires extreme caution and its own strict rules. However, an elementary understanding of what genetic engineering entails makes it clear that these claims are absurd. In traditional forms of breeding, variety is achieved by selecting from genetic traits already existing within a species' gene pool. Breeders can cross varieties, or even species, that wouldn't necessarily reproduce if left to themselves, but nature imposes strict limits to this practice. A rose can cross with a different kind of rose, but not with a mammal. A horse can reproduce with a donkey, but not with a caterpillar. Even when different species that are closely related succeed in reproducing (such as a horse and a donkey), the offspring (such as a mule) are usually infertile, thereby prohibiting the newly created organism from reproducing itself. These natural reproductive boundaries are essential to the integrity of any species.

In contrast to traditional breeding, genetic engineering involves identifying the genes that code for a specific trait or characteristic in a species, and transferring those genes into another species, often across entire taxonomic phyla or kingdoms. For example, a gene or sequence of genes in a species of arctic fish (such as flounder) that leads to the production of a chemical with anti-freeze properties can be spliced into the genetic material of a tomato or strawberry, in order to make it frost-resistant. It is now possible for scientists to introduce genes taken from bacteria, viruses, insects, animals or even humans into plants. This new technology manipulates organisms in a fundamentally different and hazardous new way, precisely because it allows us to transcend the reproductive limitations imposed by nature, and because its anticipated effects can never be completely certain.

Genetically engineered foods already on the market in the US include corn, soybeans, potatoes, squash, tomatoes, chicory and papaya, as well as milk and other dairy products from cows treated with a genetically engineered growth hormone (rBST). BST (also known as rBGH—recombinant bovine growth hormone) is a genetically engineered hormone injected into one-third of dairy cows in the US to increase milk production. It is currently banned in Europe. There is a real possibility that the US Government and Monsanto, the manufacturer of BST, will use the WTO to force the product into the EU. The BST hormone causes a five-fold increase in the protein IGF-I, which passes into the milk. An EU Scientific Committee report links IGF-I to breast and prostate cancer. It causes increased infection and disease in cows, making them produce more pus, and causing a substantial increase in mastitis, sores, foot problems, and reproductive disorders. These problems in turn increase the use of antibiotics, which of course end up in the milk.[21]

There is growing evidence that genetic engineering poses new risks to ecosystems, with the potential to threaten biodiversity, wildlife and truly sustainable forms of agriculture.[22] Critics of the technology argue that once

GMOs have been released into the environment, they may transfer their characteristics to other organisms in a way that is uncontainable. A study by the National Pollen Research Unit shows that the wind can carry viable maize pollen hundreds of kilometres in 24 hours.[23] Current trial plots where genetically modified crops are grown have a 'buffer zone' of only 200 metres between them and non-modified crops of the same species. Genetically modified crops can cross-pollinate with other crops and wild relatives, and pass on their resistance to weedkillers or viruses. The offspring may become persistent weeds within arable fields. Ever more chemicals will be needed to control the problem.[24]

Since the original publication of this book, many of the concerns of biotechnology critics have been proven justified, both inside and outside the laboratory. For example, one study showed that ladybirds feeding upon aphids that had eaten genetically modified potatoes lived half as long and laid 38 per cent fewer eggs, which were four times more likely to be unfertilised and three times less likely to hatch.[25] Another study, at Cornell University in the US, revealed that monarch butterfly larvae feeding upon milkweed that had been sprinkled with genetically modified corn pollen were significantly impaired in their development, with a large increase in premature death.[26] In several countries where genetically modified crops are grown, farmers who do not use modified seed have reported contamination by pollen from nearby modified crops, with their own crops eventually exhibiting the GMO characteristics.

The food biotech industry is dominated by a handful of multinational corporations holding interests in food, additives, pharmaceuticals, chemicals and seeds. These corporations are beginning to monopolise the global market for GMOs. This process is being facilitated through the WTO, which makes it difficult for countries to refuse a new product or technology, even if they have concerns about its potential impact on health or the environment. Patenting rights allow corporations to patent new genetically engineered varieties. This gives them control over huge areas of the market. The high cost of researching, developing and patenting new crops means that larger companies are likely to continue to dominate the market. This process is also being assisted by a planned programme of acquisitions and mergers, incorporating seed companies, genetic engineering companies and other related interests. Monsanto, for example, spent $8 billion on new acquisitions between 1996 and 1998. According to Robert T. Fraley, co-president of Monsanto's agricultural sector, 'This is not just a consolidation of seed companies; it's really a consolidation of the entire food chain.'[27]

Part of the current trade agenda is the forging of an agreement on trade in GMOs. At issue is the right of governments in Europe and elsewhere to respond to the wishes of the general populace and ban or restrict the importation of genetically modified food. The US government, at the behest of agribusinesses, grain traders and seed corporations, has fought rigorously against any such restrictions, arguing that there is no proven scientific basis for doubting the safety of these food products, and therefore, no WTO-legal basis for restricting their importation. Leaving aside the claims of 'no proven scientific basis', it is disturbing that in this process, democracy seems to have been thrown out of the negotiations altogether—that the will of a large majority of citizens does not seem to matter.

## A Shift in Consciousness: Resistance . . .

In response to the worrying developments described above, citizen groups around the world are beginning to realise that it is the corporate-dominated global economic system that is the prime culprit behind many of today's social and environmental problems. Increasingly, activists and citizen groups of all types are joining together and pressing for major policy change at both national and international levels, in order to take control of the global financial markets that are systematically subordinating human and environmental well being to a volatile, speculative market.

A very visible example of growing activism has been the aforementioned massive public resistance in many European countries against the genetic modification of foods. In the face of biotech multinationals and the US government trying to force genetically modified foods down the throats of European consumers, public pressure to restrict or even ban imports of these foods has been escalating. As a consequence, it has become impossible for European governments to ignore their voters. In the name of sovereignty and consumers' rights, some of these governments even seem willing to risk a trade war with the US, and the breakdown of WTO talks in Seattle was at least partly due to differences on this issue. In addition, the four major supermarket chains in Britain, and several others on the European continent, have publicly stated that they will not allow genetically modified ingredients to be used in their own brands. The US-based snack food giant, *Frito-Lay*, recently announced that it would not use genetically modified corn in its snack chips. In Montreal, in January 2000, agreement was reached on the Cartegena Protocol on Biosafety, establishing an international regulatory regime based on the 'precautionary principle' to manage the unique risks of the handling, transfer and use of genetically modified organisms. This means that if there is any doubt as to the absolute safety of a GMO, governments should take the precaution of keeping it off the market, rather than forcing it onto the market until proven unsafe. Although the Montreal agreement unfortunately does not provide for any means of enforcement, this acceptance of the 'precautionary principle' is a clear moral victory for activists and consumers, and it works against corporate agricultural and biotech interests, who argue that restrictions should not be imposed, unless justified by 'sound science'. All of these positive developments were the result of citizen groups and individual consumers making their opinions known.

A previous, and less publicised, victory for citizens was the stalling of the Multilateral Agreement on Investment. The MAI was an international agreement, written mainly by representatives of transnational banks and corporations and government trade officials, which aimed to force governments to relinquish much of their power, especially their ability to protect their citizens and maintain social, environmental and health standards. A relatively small number of activists and informed citizens put pressure on governments around the world and forced the stalling of the agreement—a feat made even more impressive by the fact that these negotiations were conducted in secrecy. Most elected officials, even ministers, were unaware of the existence of the MAI!

A third example is the US Department of Agriculture's retreat from its plans in 1998 to weaken organic standards, in order to allow large agribusinesses to take advantage of the increasingly lucrative market for organically grown foods. The new guidelines would have even allowed for the inclusion of irradiated, genetically modified, and toxic-sludge-treated foods under the 'organic' label. After the USDA offices were flooded with thousands of irate letters from consumers, the department backed away from its plans. The USDA has recently issued a new set of proposed organic guidelines, which generally reflects the values of organic food advocates and incorporates most of the consumers' wishes. However, big business still threatens to water down and distort organic standards.

There is much citizen resistance taking place in the South. In India, where action against the global economy is snowballing, a 'Declaration of the Indian People Against the WTO' has recently been drawn up by a group of farmers, academics, NGOs, unions and others. It attacks the 'WTO-IMF-World Bank trinity', which 'has the potential not only to suck the sweat and blood of the masses of two-thirds of the world, but has started destroying our natural habitats and our cultural diversity'.

Also in India, the ten-million-strong Karnataka state farmers association (KRRS) has been involved in destroying Kentucky Fried Chicken outlets, occupying the offices of the giant seed company Cargill and, most recently, burning Monsanto's genetically modified crops, all in the wider context of protesting about the effects of globalisation on the subcontinent. The KRRS has also organised an 'Intercontinental Caravan' of Indian farmers and others, to tour the world, raising awareness about the problems of globalisation.

But undoubtedly the most encouraging and significant expression of resistance was the mass protest in Seattle. As mentioned earlier, the demonstrations there involved an extraordinary array of farmers, business people, mothers with young children, environmentalists, indigenous people and members of labour unions. Protesters numbered in the tens of thousands, and brought world-wide attention to a process that has over the years taken place in total secrecy. The message of the people marching in the streets was very clear: globalisation is not a natural or evolutionary process; it is about specific trade agreements and government policies, and these must be changed. The atmosphere of resistance created by these protests undoubtedly played a major role in the collapse of these talks, and they have certainly ensured that future trade decisions—which so fundamentally affect the well-being of the planet and its citizens—will no longer be made outside the glare of public scrutiny.

These successes truly signal a dramatic change in public awareness. The *Economist* article discussed above noted that 'the critics are not just against the WTO, they are against globalisation broadly defined'. Ordinary people are becoming aware that 'governments are supporting corporations to the detriment of jobs and the environment'.[28] As a result, the warning to business and governments is clear: 'Globalisation is not irreversible'.[29] The whole dogma of 'free' trade has been based on the notion that deregulation increases freedom (to trade). Increasingly apparent, however, is the reality that 'deregulation' actually involves the imposition of rules friendly to big business, in place of the hard-won environmental and human rights protections instituted by national governments under pressure from the public. Never before has there been such public resistance and awareness when free trade treaties have been negotiated. For the first time, there is real hope that the madness of economic globalisation can be brought to a halt.

## ... And Renewal: the Local Food Movement

Alongside this widespread, and increasingly effective, resistance to the corporate-led global economy, a plethora of positive micro-trends at the grassroots level is demonstrating the alternative to globalisation: localisation. Probably the strongest and most significant aspect of this trend is the local food movement. It is now overwhelmingly clear that globalisation is destroying local food production. As described above, in this process, food security is threatened, the nutritional quality of food is eroded, pollution increases, and biodiversity, rural landscapes and regional distinctiveness in diet are homogenised. The genetic engineering of new food varieties threatens to take us further down this destructive path. In response to these developments, the local food movement—local production for local consumption—has become one of the most successful grassroots examples of economic localisation in action.

### *Local Food Production*

The benefits of local food production are many. Growing for local markets makes small-scale organic agriculture much more viable, and the diversity encouraged by local marketing obviates the need for toxic chemicals and other off-farm inputs. Crops are less susceptible to blight because diversified farms are more resilient. If disease does strike, it will affect only a relatively small portion of the farm's total output. Pests are less of a problem because the agricultural environment encourages the presence of natural predators. Weeds are less of a hindrance because intercropping—growing different crops in the same space—naturally helps to suppress them, and there is less space between crops for them to flourish. What weeds there are can either be mechanically suppressed by hand pulling or fire, or simply tolerated. (Many so-called 'weeds' have valuable uses in their own right, and the term 'weed' is usually used to refer to any plant that interferes with the growth of monocultures for the market, irrespective of that plant's usefulness as a source of food or medicine). The decreased dependence upon inputs dramatically reduces the environmental impact of food production. Non-renewable natural resources are conserved, with farmers instead using the resources that abound on the farm itself, many of which are considered 'wastes' in industrial food production systems, and soils and watercourses are not poisoned. The smaller scale obviates the need for large, expensive farm machinery, and thereby allows the incorporation of trees and hedges into the agricultural landscape, providing a valuable habitat for birds and animals, and improving the aesthetic quality of the rural environment.

Contrary to common belief, small-scale, diversified food production is also better at meeting people's needs. The orthodox view of agricultural development holds that industrial agriculture—with its consolidation of land, mechanisation and use of chemicals—has vastly increased agricultural productivity. In fact, however, the productivity of industrial agriculture is largely a myth, propagated for years by the beneficiaries of this type of agriculture. Study after study, carried out in diverse locations all over the world, shows that small-scale, diversified agricultural systems almost always have a higher *total* output per unit of land than large-scale monocultures.[30]

## *Local Food Marketing*

Faced with a wasteful and increasingly hazardous food culture, people are beginning to appreciate the advantages of local food, and are working to rejuvenate markets to obtain it. A variety of proven local marketing methods are currently being tailored to suit the needs of different communities:

FARMERS' MARKETS. Farmers' markets were once common throughout the world, but experienced a major decline when large-scale monocultural production made local marketing economically unviable. Now, however, there is a resurgence of these markets. In the UK, the first farmers' market in recent years, in the city of Bath, generated much public interest, leading to a wave of new markets throughout the country. Since the early 1990s, the UK went from having no farmers' markets at all, to having over 140.[31] Farmers' markets are growing rapidly in North America as well. The US Department of Agriculture's directory listed 2,675 farmers' markets in 1998, over 50 per cent more than there were just four years earlier.[32]

Farmers' markets are sometimes held in a permanent structure, but more often occupy a section of street closed off to traffic. Most of the producers selling at these markets are small-scale, and even if they are not certified organic, they tend to use fewer pesticides than large-scale monocultures. Consumers have a range of in-season fresh produce to choose from, often harvested that same day. At most markets, they can also choose from a variety of other products, such as honey, fresh jams and preserves, ciders and juices, bread, cut flowers and potted plants. Many have vendors selling a variety of meats and dairy products, most of which are free of the hormones and antibiotics found in goods that come from large livestock enterprises. Some farmers' markets even feature live music, played by either paid performers or street musicians. All of these markets promote a sense of community, both through the friendly, relaxed shopping environment they create, and through the fact that money spent there supports local enterprises, and remains within the community.

COMMUNITY SUPPORTED AGRICULTURE (CSAS). CSAs are the next best thing to growing your own food. The basic model is simple: members pay producers for a certain number of weekly deliveries in advance, and every week, the boxes of produce are delivered, either to the subscribers' homes, or to a central pick-up point. The benefits are obvious: consumers help small-scale, local farmers by providing a guaranteed market for their produce, and, in return, growers provide a weekly share of fresh food for a relatively low price. Most offer a mix of eight to twelve vegetables, fruits and herbs per week; some link up with other CSAs to maintain diversity; and others offer value-added products such as cheese, honey and bread. As they are growing for actual people, rather than for an abstract market, farmers want to provide a reasonable variety of produce, which thereby gives them an incentive to diversity their production, which in turn encourages them to employ more ecological practices.

CSAs not only enable consumers to establish a personal relationship with the person who grows their food, they also provide an opportunity for urban consumers to reconnect with the land. With most CSAs, the farmer welcomes visits from subscribers, perhaps even organising 'workdays' that enable subscribers occasionally to help out on the farm. Most important, consumers have the benefit of being able to eat produce that is much fresher and more nutritious than nearly anything they could buy at the supermarket. Members know where their food originates and how it was produced, and farmers receive payment in advance, rather than when the harvest is in. In this way, trust and mutual support are restored to the vital relationship between farmers and consumers.

There are currently some 600 CSA operations in the USA, involving 100,000 consumers, and a turnover of $10–20 million per year.[33] In the UK, there are now nearly 200 produce box schemes, selling to more than 45,000 households, and representing sales of £22 million.[34] A local food-promoting scheme in the Forest of Dean, begun in the late 1990s, sold nearly £325,000-worth of local food to local people in its first year. The 'Forest Food Directory' lists 32 different food producers, with products ranging from organic and free-range meat, through to vege-box schemes and local cheeses.[35]

A recent study of 83 CSAs in the USA, carried out by Timothy Laird and reported by Richard Douthwaite, has shown what people most value from them. More than 60 per cent of the farmers said that the most successful aspect of their operations was the strengthened bonds between people, resulting in networks that: 'reconnected people with the land and reconnected farmers with the people who eat the food that they grow'. Only about one-third of farmers mentioned the value of financial stability.[36]

People buying food direct from the producers are often very enthusiastic about the quality, and about the manner in which it is bought. In her book *Local Harvest,* Kate de Selincourt quotes a satisfied vege-box customer: 'My husband says it's like having Christmas every week. The quality is superb, it's sort of a lucky bag . . . the kids rush to see what we've got. There is no possible comparison with the taste. You feel really sorry for people going to the supermarket.'[37]

Farmers are also satisfied with such direct relationships: when farmers are allowed to sell in the local marketplace, more of the profit stays in their hands. Currently, when the consumer buys food for £1 in the supermarket, the farmer very often receives less than 5 pence. The consumer spends 95 pence of that pound on transport, packaging, irradiation, colouring, advertising and corporate profit margins. But when these links are closed, the farmer receives more money and the consumer pays less. Both win. Kate de Selincourt asked farmer Pat Finn why she sells direct to customers rather than through a supermarket or butcher's shop: 'We can't provide what they call "portion control": everything the same size every time. . . . We really enjoy the personal side of the work—it is nice to think that we have become so friendly with people just through business.'[38]

The joy of a direct connection between producers and consumers is often that their ideals coincide. They want the same things: human-scaled and human-paced production with high quality. They both want freshness, variety and a fair price. Both social life and business often flourish when suppliers and consumers form face-to-face relationships.

### *The Promise of Local Food*
These local food systems are not an idealistic dream: they are a reality, now, and they are showing that they can be replicated across the world. They do not need to grow bigger: the farmer does not need to produce more and more to flourish, does not need to sell at ever greater distances. These systems are proliferating everywhere. They are taking the economy from global to local, and proving that it can be done.

But for such local food systems genuinely to flourish and multiply, changes at policy level are clearly necessary. The current trade and finance policies described earlier are artificially lowering the prices of industrially produced foods by shifting the costs of production on to the community, forcing citizens to pay for the global trade infrastructures and subsidies to big business that industrial food requires. If groups campaigning for sustainable farming, wildlife issues and better food do not take these hidden subsidies into account, and do not challenge the economic basis of our current monocultural, export-based food system, they risk falling into the trap of arguing that consumers should pay more for better food—when, as farmers' markets show, they can actually pay less. The industrial approach marginalises the poor, who often cannot afford healthy, high-quality food. Furthermore, to overlook hidden subsidies is to miss a fantastic opportunity: if these resources were diverted towards decent agriculture and retailing, society would produce better food at no extra cost at all.

Recognising the consequences of current economic policies also gives agricultural and environmental groups common cause with those campaigning for social justice and human rights internationally. As clearly demonstrated by the experience of Seattle, these diverse bodies are now beginning to join hands to demand a different set of policies, and the redrawing of the global economic map.

The most important thing to remember is that we do have the power to change things. The destructive global economy can exist only as long as we are prepared to accept and subsidise it. We can reject it. And we can start by enjoying the wealth of benefits from re-linking farmers and consumers. Fresh, local food for all may be one of the most strategic—and certainly the most delicious—ways to reverse economic globalisation.

## Notes

1. *The Economist*, vol. 353, no. 8147, 27 November 1999.

2. Debi Barker and Jerry Mánder, *Invisible Government: The World Trade Organization: Global Government for the New Millennium?*, San Francisco: International Forum on Globalization, 1999, p. 22.

3. Department of Environment statistics, cited in H. Raven and T. Lang. *Off Our Trolleys? Food Retailing and the Hypermarket Economy*, London: IPPR, 1995.

4. Ministry of Agriculture, Fisheries and Food, *Agriculture in the United Kingdom, 1997*, London: MAFF, 1998. According to SUSTAIN, in 1997, the UK imported 126 million litres of liquid milk, while exporting 270 million litres; imported 23,000 tonnes of milk powder, while exporting 153,000 tonnes, 135,000 tonnes of which went outside the EU. In

that year, the UK also imported 115,000 tonnes of butter, 51,000 tonnes of which was from outside the EU, and exported 67,000 tonnes of butter, 27,000 of which was exported outside the EU. In 1996, the UK imported 434,000 tonnes of apples, 202,000 tonnes of which came from outside the EU. Over 60 per cent of UK apple orchards have been lost since 1970. In 1997, the UK imported 105,000 tonnes of pears, 72,000 tonnes of which were from outside the EU. 'Food Miles—Still on the Road to Ruin? An assessment of the debate over the unnecessary transport of food, five years on from the food miles report', Sustain—the alliance for better food and farming, October 1999.

5. Herman Daly, 'The Perils of Free Trade', *Scientific American*, vol. 269, no. 5, November 1993, p. 51.

6. J. Koojiman, 'Environmental assessment of packaging: sense and sensibility', *Environmental Management*, vol. 17 no. 5, New York: Springer-Verlag, 1993; SAFE Alliance, *Food miles report*, London: SAFE Alliance, 1996.

7. Anon., 'Do you need all that packaging?', *Which?*, London: Consumers' Association, November 1993, p. 5.

8. Non-biodegradable packaging is not the only waste that goes into landfills, and we are now squandering a valuable resource in our failure to utilise organic waste. A quarter of London's waste (900,000 tonnes per year) is green and putrescible, much of it from food (T. Garnett, *Growing Food in Cities: a report to highlight and promote the benefits of urban agriculture in the UK*, London: SAFE Alliance and National Food Alliance, 1996). For millennia, this organic waste has been an important resource for farmers, either as compost or as food for livestock. Now it is a liability, which must be trucked away and dumped into dense landfills.

9. Environmental Research Foundation, 'Incineration News', *Rachel's Environment & Health Weekly*, no. 592, 2 April 1998.

10. Robert C. Taylor, Professor of Agriculture and Public Policy at Auburn University, in testimony before the US Senate Agriculture Committee hearings on concentration in agribusiness, January 1999. Cited in A. V. Krebs, *The Agribusiness Examiner*, no. 57, 23 November 1999.

11. Krebs, op. cit.

12. Retailing Enquiry, Mintel, March 1996.

13. D. Hughes, 'Dancing with an elephant: building partnerships with multiples', paper presented at The Guild of Food Writers' conference, 'The Vegetable Challenge', London, 21 May 1996.

14. From an FAO study, based on 150 country reports, *State of the World's Plant Genetic Resources*, Rome: FAO, 1996.

15. Jack Doyle, *Altered Harvest*, New York: Viking, 1985; FAO statistics, www.fao.org.

16. IMF Statistics Department, *International Financial Statistics*, February 1999, vol. 53, no. 2, Washington, D.C.: International Monetary Fund, 1999.

17. Hoover's Inc., *Hoover's Handbook of World Business, 1997*, Austin, Tex: Hoover's Business Press, 1997; *The World Economic Factbook*, 4th ed., London: Euromonitor, 1996.

18. C. LeQuesne, Reforming the World Trade Organisation. The Social and Environmental Priorities, Oxford: Oxfam, 1996.

19. American Soybean Association, 'European Response to Genetically Modified Soybeans', November 1996.

20. Luke Anderson, 'Genetically Engineered Food, No. 5', Greenpeace, www.greenpeace.co.uk.

21. Ibid.

22. See R. Steinbrecher and Mae-Wan Ho, 'Fatal Flaws in Food Safety Assessment: Critique of The Joint FAO/WHO Biotechnology and Food Safety Report', 1996, www.psagef.org/fao96.htm.

23. Science and Technology Select Committee Hearing, 26 April 1999, quoted in Greenpeace UK and Soil Association, 'The True Cost of Food', London: Greenpeace UK, June 1999.

24. Greenpeace UK and Soil Association, op. cit.

25. Science and Technology Select Committee Hearing, 26 April 1999, op. cit.

26. John E. Losey, Linda S. Rayor and Maureen E. Carter, 'Transgenic pollen harms monarch larvae', *Nature*, vol. 399 no. 6733, 20 May 1999, p. 214.

27. Quoted in J. Flint, 'Agricultural industry giants moving towards genetic monopolism', Telepolis, Heise Online, 1998.

28. *The Economist*, op. cit.

29. Ibid.

30. Peter Rosset, 'The Multiple Functions and Benefits of Small Farm Agriculture in the Context of Global Trade Negotiations', *Policy Brief* No. 4, Oakland, Calif.: Institute for Food and Development Policy, September 1999; R. Albert Berry and William R. Cline, *Agrarian Structure and Productivity in Developing Countries*, Baltimore: Johns Hopkins University Press, 1979; Gershon Feder, 'The Relationship between Farm Size and Farm Productivity', *Journal of Development Economics*, vol. 18, 1985, pp. 297–313.

31. Soil Association (UK), Local Food Links campaigner, personal correspondence.

32. United States Department of Agriculture, National Directory of Farmers Markets, Washington, D.C.: USDA, Agricultural Marketing Service, November 1998.

33. This assumes an average share of $550, an average number of members of 50, and 600 CSAs.

34. Soil Association, *The Organic Food and Farming Report, 1999*, Bristol, UK: Soil Association, 1999, p. 25.

35. M. Dunwell and K. de Selincourt, *Forest Food Directory: A Guide to Local Food Production in the Forest of Dean*, Cheltenham, UK: Vision 21, 1997.

36. T. Laird (1995), reported in Richard Douthwaite, *Short Circuit: Strengthening Local Economies for Security in an Unstable World*. Totnes, UK: Green Books, 1996.

37. Kate de Selincourt, *Local Harvest*, London: Lawrence & Wishart, 1997, p. 164.

38. Ibid., p. 167.

# The Hamburger Bacteria    19

NICOLS FOX

*Nicols Fox is a journalist whose work has appeared in* The Economist, *the* New York Times, *the* Washington Post, Lear's, Newsweek, *the* Boston Globe, USA Today, *the* Christian Science Monitor, American Journalism Review, *the* Los Angeles Times, *the* Atlanta Journal-Constitution, Columbia Journalism Review, The Washingtonian, Art in America, *and many other publications. She has appeared on numerous radio and television programs as an expert on foodborne illness, including* 48 Hours *and* Today, *and on* Nightline, *debating with the U.S. secretary of agriculture.*

*In this essay, Fox explains the dangers of* E. coli O157:H7, *a new deadly form of the* E. coli *bacteria that has sometimes infected meat and organic food in North America.*

> *The microcosm is still evolving around us and within us.*
>
> Lynn Margulis and Dorion Sagan, *Microcosmos*, 1986

IN FEBRUARY 1992 Mary Heersink, the wife of an Alabama physician and mother of four, was driving two of her sons to Florida. It was a wonderful, balmy Saturday in early spring, and she'd taken the boys out of school so that they could do something special.

Damion, eleven at the time, had returned from a Boy Scout camping trip the weekend before and was looking forward to attending space camp. His younger brother, Sebastian, was headed for a tennis camp. Damion, she remembers now, didn't seem himself. He was pale and quiet. She thought he might be coming down with flu, and if it developed further, she thought she

---

From *Spoiled: Why Our Food Is Making Us Sick and What We Can Do about It* by Nicols Fox. Copyright © 1998 by Nicols Fox. Reprinted by permission of Basic Books, a member of Perseus Books, L.L.C.

would keep him in the hotel room with her for a day or so until he recovered, then send him on to camp. "He was very groggy and hard to awaken. He seemed mentally detached," she says now.[1]

Along the way he became ill with stomach cramps, then fierce diarrhea. She wasn't overly worried at first. Then it became bloody diarrhea. In fact, it was more blood than diarrhea, she says. "It looked like pure clotted blood, and it just came constantly. At that point you know you have to get to a hospital."

She made it to Tampa, her destination, quickly checked into the hotel facility adjacent to the tennis camp, where she left Sebastian, and took Damion to the University Community Hospital. The name reassured her. An association with a university was a good sign.

Mary Heersink had managed, while checking in, to contact her husband Marnix, an ophthalmologist, who was attending a continuing education course on pediatric medicine in Birmingham that weekend. Discussing the symptoms over the phone, they concluded it was just a vicious case of the "flu." They assumed that a trip to the hospital could secure some kind of medication that would make Damion comfortable, then Mary would turn around and get him home.

It wouldn't work out that way.

The hospital was busy and crowded. "It was like a war zone in there," she says of the emergency room. Damion's problem, acute diarrhea, didn't seem that compelling. He wouldn't be seen before 10:30 that night, five hours after they had come in. Within hours they were getting the results of his tests. "They were extremely alarming," she remembers. "His body was destroying blood platelets [small disks, smaller than cells, that promote coagulation] and shredding red blood cells." A gastro-graph showed that his intestine was edematous—"very puffy, very abnormal-looking." He was hospitalized at that point, but his condition worsened rapidly. "He went from bloody diarrhea to hallucinating, platelets being destroyed, kidneys malfunctioning. He was just plummeting," she says.

It would take an astute gastroenterologist to diagnose his condition as hemolytic uremic syndrome; this doctor insisted that Damion be transferred to St. Joseph's Children's Hospital in Tampa. (University Community Hospital had no connection at all to any university, Mary would later discover.) That in itself took hours. The confusion was almost intolerably frustrating. "There were mix-ups. They kept bumping my son. And it was so urgent," Mary remembers.

Neither she nor her physician husband had heard of hemolytic uremic syndrome. Still in Birmingham, Marnix Heersink tried looking it up in the library of the medical school library and found almost nothing. Both he and Mary were misled by the term "syndrome." It sounded better than some of the other possible illnesses that had been suggested. She would subsequently find out how horrifying HUS is—an illness that would attack first one then another of the chief organs in Damion's body in what is described as a cascading pattern.

The doctors had assembled a team to deal with his case. As gently as they could, they told Mary what the mortality rates were and what she could expect. She was told that the course of the disease might well be five or six weeks; that it would be like a horrible roller-coaster ride as the disease created an anarchical situation in the body. There would be swelling of each organ affected. There would be clots. There could be strokes. Mary relayed this to her husband over the phone, and he booked a plane.

Marnix Heersink, born in Holland, is tall, handsome, and scholarly, with a ready smile that belies his otherwise serious demeanor. The name of his son's syndrome had told him one thing—it was as much a blood disease (hemolytic) as a kidney (uremic) disease. He thought of a fellow student at Western Ontario, where he'd done his medical training, who'd gone on to specialize in blood diseases, and although it was a wild shot, he contacted his friend's office in Canada. He explained to the nurse why he was calling, but when told that his friend was on a two-week vacation out of the country, he didn't bother to leave a number.

Damion was now strapped to his bed in intensive care and on kidney dialysis. A peritoneal catheter had been inserted in a surgical procedure into his abdominal cavity to take on the work of the failing kidneys. He could have no food or drink, and no tranquilizers that might disguise his neurological condition. His platelet count had dropped to precipitous levels. He might hemorrhage internally, they were told. Three days before he had been a strong, healthy, active boy. The Heersinks were frantic.

No one knows how many people get HUS each year, but estimates are that the disease kills as many as five hundred, most of them children. If the mortality rate is figured at around 5 percent, that means there must be at least eight thousand cases, but the CDC assumes that many go undiagnosed. What was an obscure disease just fifteen years ago is now the leading cause of renal failure in U.S. and Canadian children. And its numbers are increasing.

The doctors at St. Joseph's suspected at once where it had come from. Damion knew, too. He was in and out of consciousness in the early days of his illness, but in a moment of lucidity he told his father, "I know what made me sick."

On the Boy Scout camping trip a few days earlier, one of the projects of the group of ten-, eleven-, and twelve-year olds had been the preparation of food. Hamburger was on the menu. Damion, who had little practical experience with meat because his health-conscious family seldom ate it, had formed the patties and a corner of one had fallen off. While the burgers were cooking, this small piece of ground beef was not. The morsel lay on the platter, exposed to the air and discoloring. When the burgers were returned to the tray, it was beside them, grayish-brown and looking for all the world like a piece of cooked meat to Damion. He popped it into his mouth. He realized his mistake at once, but not wanting to look foolish in front of his friends, he swallowed it quickly. In all likelihood that indeed was what made him sick.

Hamburger meat is the chief culprit in infections caused by the bacterium *E. coli* O157:H7. Ground meat is especially vulnerable because the grinding spreads the bacteria, a product of fecal contamination of meat, throughout the product. The most unfortunate of those who begin with the bloody diarrhea that characterizes the infection will develop HUS. Researchers think that 85 percent of all HUS cases in the United States are caused by *E. coli* O157:H7.

As Damion lay ill, growing worse by the moment, Dr. John Kelton, Marnix Heersink's friend, called his office at McMaster University for his messages and learned that his friend's son was in trouble. He had no idea where the Heersinks were. By calling friends and relatives, he was able to track them down. He told them, to their astonishment, that he was on his first vacation in years, in Florida, and that HUS was his specialty. He would come on board—via teleconferencing and the fax machine—from his vacation site on the eastern coast.

Kelton suggested an aggressive treatment of the blood using plasma exchange. Used in France and Canada with good results, the technique is less frequently employed in the United States, where nephrologists have claimed HUS as their disease. It's a sophisticated and costly treatment, and the St. Joseph's team had some reservations. But Kelton persuaded them of its effectiveness. And Damion was getting worse. It was clear they had to do something.

The boy was infused with eight units of fresh frozen plasma, which was exchanged for his own plasma. "What was coming out was just evil-looking," says Mary. "Foul and viscous-looking, discolored; full of shredded blood cells and toxins, which I couldn't see, of course, but were there. It just looked bad. And there was a strong, alien, most offensive odor. It was coming from everywhere; his hair, his skin, everything."

Damion underwent eleven plasma exchanges over the next two to three weeks, each lasting two to three hours each. After the exchanges, she noticed that the smell disappeared. A period of unwarranted optimism can occur at this point. Damion was suddenly improving. But the toxins had already done their damage and complications could set in. They could kill.

What would follow was a horrifying series of events. Damion's heart enlarged. His lungs filled with fluid, and he had to go on a respirator. Three times he underwent pericardialcentesis: With Damion awake under a light local anesthetic, the doctor inserted a needle just inside the pericardium, the sac enclosing the heart, to drain off the fluid. Each time a liter of fluid was drawn off. Then it would begin again. His heart rhythms began to be affected. At one point "he became convulsive. He was turning blue. His blood pressure plummeted to 40/20. He was dying. We had to leave the room, and they brought in the crash cart," says Mary.

The doctors patched him up and told her they would have to use more drastic surgical procedures to drain off the fluid. The idea was to cut a little window in the pericardium to release the fluid. When the surgeons got inside, they found that the pericardium was in such bad shape, she says, "so nasty-looking, so ragged," that "it had to be stripped off and discarded." No one really knows what the long-term implications are of having no pericardium, but the doctors told the Heersinks that any number of people were walking around without them. Damion could have a normal life, they said. They didn't have to add: if he survived.

Even then, Damion's course would not be smooth. The first sip of fluid by mouth was followed by agonizing pain. He had perforated his intestines and would have to go back to surgery. This time he was "split," as his mother says, "from neck to pubis, the way you eviscerate a cow." They found a small hole that unfortunately had spilled its contents into the lower body cavity. The doctors washed Damion's intestines and packed them back into his ravaged body. Somehow he would avoid massive infection, but there were abscess pockets in his abdomen, and a few days later he would go back into surgery. Incredibly, after a total of seven surgeries, Damion finally began a genuine recovery. And this time there would be no further setbacks. He had lost twenty-five pounds and had "absolutely no muscle tissue." When he was finally able to sit up on the bed, Mary "cried and cried to see the back of him look like that, so bony, so frail."

The mortality rate for children in his category of HUS is 5–10 percent. If a child has a complicated case, as Damion did, there is a one-in-five chance of death and a 74 percent chance of a major physical impairment. Children who have survived are often left with "souvenirs" such as diabetes, blindness, a colostomy, or kidney failure that requires dialysis or a new kidney. Damion was lucky, from beginning to end. His intestines had to relearn how to process food, and he had to learn to stand again, but finally, on his twelfth birthday, April 11, after seven weeks in intensive care, he was allowed to leave the hospital. "Then we spent a year putting him back together," says Mary.

Time after time during the ordeal she'd thought that her oldest son, her precious fair-haired boy, might be taken from her. Now he was home, recuperating slowly, but while caring for him, getting his body to work again, she found that the questions wouldn't go away. Her doctor father and brother-in-law, both ophthalmologists like Marnix, had bombarded her with questions when Damion was in the hospital. "What is the etiology of this syndrome?" they had sensibly wanted to know. What, in other words, was the cause? The members of this family were not merely inquisitive, they were confident people who demanded answers and expected to get them. The disease made no sense in the context of twentieth-century hygiene. Where had it come from? She remembered that Dr. Kelton's first question when he heard of Damion's illness was, "So when did he eat the hamburger?"

She was puzzled. There is hardly an American who does not think of hamburger as the National Food. No one had ever said ground meat did anything more ominous than raise cholesterol. Her brother-in-law had sent her articles from medical journals about HUS while Damion was still in the hospital, and she'd read about the hamburger connection again—and about a bacterium with an impossibly long name. What was it? Why was it in meat? Why had no one ever heard of it before?

At home, when she could spare a few minutes away from Damion and the rest of her family, she'd gone to her local medical center library and done some research. One name stood out as she went through the documents: P. Griffin at the CDC. Dr. Patricia Griffin had made the new bacterium her specialty. Boldly, Mary picked up the phone and called her. And Dr. Griffin listened. What Mary had already learned surprised her. Where there were gaps in her new knowledge, Griffin filled in the blanks. Coupled with what Mary already knew, the larger picture was horrifying. What was the matter with meat inspection that it allowed meat to be sold that was capable of putting people through what Damion had been through? Why wasn't anybody doing anything about it?

The questions didn't go away. Even as Damion improved, when she wasn't driving her three other children to their soccer games and school functions, she accumulated more information. She read reports from the General Accounting Office about the weaknesses of meat inspection procedures; she contacted a consumer group called the Government Accountability Project that sent her reports and the testimony of whistle-blowing meat inspectors. The more she learned about meat inspection policies, the more unacceptable the situation became.

Then one day in early 1993 Marnix Heersink came home with *USA Today* tucked under his arm. "Look," he said, "it's happening again." Together they read the first reports of what would eventually be known as the Jack-in-the-Box outbreak. The American public would read with them and learn what everyone should have learned a decade before, when the story actually began.

## American Nightmare

Vegetarians of longstanding, who long ago stopped dreaming of steak and roast beef, tell me that in weak moments, with their defenses down, the memory of hamburger still returns to torment and test their will. The marriage of soft roll to juicy meat, adorned with the perfect blend of condiments, hot, dripping, and flavorful, has prospered and endured and proliferated beyond anyone's imagining. From its uncertain origins in the 1800s—two restaurateurs lay claim to its invention—the hamburger has become the most American of foods. But far more than that, it has become a cultural icon, a symbol of the national character. The hamburger is the backyard barbecue, the camping trip, the kitchen standby. Especially in its fast-food form, it is the American dream personified: modest, predictable, preseasoned, prewrapped, standardized, quality-controlled, and ready-to-go. It is more than a meal. It is a food wrapped in political and cultural mystique.

It was probably J. Walter Anderson, a short-order cook and future founder of the White Castle hamburger chain, who thought to place a ground beef patty not between bread, as had already been done, but between buns. Inspired, he added shredded onions. The bun, popped over the cooking burger, absorbed the flavors of the meat and onions. It was a hit. It would be refined over the years, but he had created, in its deceptive simplicity, the culinary equivalent of blue jeans. The only problem was getting the middle class to eat it. It was 1921, some fifteen years after Upton Sinclair had published *The Jungle*. That novel, with its graphic and horrifying descriptions of the conditions in slaughterhouses, had destroyed whatever illusions Americans may have had about the safety of their meat and prompted the first federal meat inspection laws. Consumer confidence had yet to be completely restored. One could cook foods carefully and thoroughly at home, but eating out was risky business. Anderson catered to this skepticism and apprehension by grinding his meat fresh and cooking his hamburgers in full view in an all-white, hospital-like atmosphere. Even the name of his establishment, White Castle, was chosen to signify purity and cleanliness, says Jeffrey Tennyson, author of *Hamburger Heaven*, a history of the burger.[2] But mothers still needed more convincing. Anderson sponsored scientific studies to demonstrate that his burgers were actually good for you.

Not long afterward, Billy Ingram, Anderson's investor, partner, and successor as White Castle president, introduced the frozen hamburger. It could be slapped, still frozen, on the grill; five critically placed holes in the frozen patty would allow the steam to escape and hasten the cooking. White Castle had invented fast food. Ingram, in a masterful triumph of public relations over reason, managed to convince customers that this dubious innovation was an advancement and an added assurance of quality. The flash freezing, the White Castle ads said, would guarantee that the beef patties were "at the peak of their perfection." Turning flaws into assets would become a standard PR approach.

The chain spawned a multitude of look- and sound-alikes, all eager to capitalize on a proven model in the growing burger market. The most successful competitor, White Tower, emphasized safety by hiring a team of "Towerettes," young women in nurselike uniforms posted for reassurance at every White Tower restaurant. By the time the multitudes of copycats had mutated into the ubiquitous drive-ins of the 1950s, with their gaudy, futuristic architecture, and then into the even faster-food franchises of the 1960s, the safety of the hamburger was taken for granted.[3]

Standing sentinel behind the reputations of the fast-food franchises was the U.S. Department of Agriculture, which inspected and approved the meat they used. Shoring up the agency's reputation was the assurance, repeated endlessly, that "America has the safest food supply in the world." From the 1950s to the 1980s that guarantee fit smoothly into the image America had of itself, an image of technological superiority, the highest standards, and an enviable lifestyle. It had become a part of the American mythology—often quoted and never questioned.

The era of the Great Hamburger Wars began in the late 1970s. Successful chains had toned down the flashy images of their recent past for something down-home and comfortable. "McDonald's and the other chains," says Tennyson, "decided to make a concerted effort to capture the family market." As a McDonald's spokesperson said in 1973, "That meant going for the kids. We decided to use television, so we created Ronald McDonald."

Kid appeal caught on, and the chains outdid themselves creating attractions for the younger set. All the while competition was growing as stronger chains gobbled up smaller ones and tough new contenders entered the arena. A fierce advertising war began in late 1981 between McDonald's, Burger King, and Wendy's. In one of the ads in the battle that followed, McDonald's described the intense urge for one of its burgers as "the Big Mac Attack." It would turn out to be an unfortunate slogan.[4]

Medford, Oregon, is a town of some forty-two thousand inhabitants in Jackson County, in the southern part of the state, just to the west of the Cascade Range. It is set in the Rogue Valley, a bowl almost, surrounded by hills, forests, and mountains, with a gentle climate that makes it ideal for growing fruit. For several decades now the area has been a draw for emigrants from California searching for a slower, more pastoral life. With the decline of lumbering, Medford has seen an influx of retirees, artists, writers, investors, and consultants who work from their homes, as well as those looking for an alternative lifestyle.

In February 1982 Dr. Richard Hebert, an internist on call at the Rogue Valley Medical Center, saw several patients with bloody diarrhea. It seemed unusual, something more than ordinary diarrhea. The cramping was fierce, like childbirth, said one female patient, and the diarrhea was often more like pure blood. It could easily be mistaken for hemorrhage. When a nurse mentioned that there were other similar cases, Dr. Hebert pulled their charts. He was struck by the similarity of their symptoms. The other area hospital, Providence Hospital and Medical Center, had patients with similar symptoms, he discovered. He also noticed that the patients had the same phone prefix in White City, a suburb of Medford. For lack of a better name, he and the staff began calling it 826 illness. Some of the patients had gone to physicians who had identified their symptoms as appendicitis. Some had even undergone surgery. No infected appendixes had been found.

Dr. Hebert had stool samples taken, and he contacted the Jackson County health officer, Dr. Taira Fukushima, who thought it might be *Yersinia enterocolitica*. When the cultures were negative for *Yersinia*, as well as *Salmonella*, *Shigella*, *Campylobacter*, and all of the several parasites that might cause such symptoms, Dr. Fukushima put in a call to the state epidemiologist's office, and from there, Dr. Steve Helgerson, an EIS officer on assignment to the Oregon Health Department, contacted the CDC. From then on, Dr. Hebert remembers, he simply "rode the wave. They were on this like white on rice," he says.[5]

Dr. Lee Riley, thirty-one years old and in his first year as an officer in the Epidemic Intelligence Service, took the call. Riley was the officer in charge of *Campylobacter* and *Salmonella*, and even though neither had been identified in the Medford cases, he suspected they eventually would be. In these early conversations with the state health department, he was struck by the fact that the victims showed little or no fever. Fever is an indicator of infectious disease, a sign that the body is doing battle with an invader. Its absence was enough to catch his attention. The health department had a number of questions, which he tried to answer, and then he made some suggestions.

The CDC heard nothing more from the Oregon Health Department for two weeks. Then it called again. There were more cases, and the department was no closer to finding a cause. Would the CDC come out and help? Riley packed his bags and got on a plane to Portland, Oregon. He linked up with Helgerson, and together they drove to Medford, discussing the case on the way. "We didn't know what was going on," Riley remembers. "We didn't even know if this was an infectious process. It could have been chemical, it could have been poison, it could have been anything."[6]

What would follow was a classic investigation. First came a "quick and dirty" case study to identify a set of hypotheses. They asked a good many questions just to find out what questions they needed to ask. This helped them to devise another set of questions that, together with the use of carefully selected controls, might lead them to the source of the disease. Although word of the outbreak was creeping out in the community, the first press coverage did not appear until March 17 in the *Medford Mail Tribune*. Allen Hallmark reported that the CDC had briefed the local board of health, but the names of the team members were not released. By the next day's edition Hallmark had found and talked to the investigators. The absence of fever, Riley told Hallmark, meant that it would be unusual if a bacterial agent were responsible.

Dr. Hebert, however, had always assumed it was an infectious disease, and gradually the cautious EIS officers were coming to the same conclusion. In some families more than one person was ill. And, Riley remembers, "One thing that kept coming up in the interviews was that the patients had eaten at one of the fast-food restaurants before they got sick." The restaurant was McDonald's. The thing was, eating at McDonald's wasn't that unusual. Given the popularity of the chain, it was almost expected. Nevertheless, they decided to include it as one of the questions in the case-control study.

In the meantime, Riley had been in constant contact with the Enteric Diseases Branch in Atlanta; as he and Helgerson were setting up the study, a team of epidemiologists and laboratory people had been just as busy looking for the cause of the outbreak in the samples. Joy Wells is the chief of epidemiological investigations for the Foodborne and Diarrheal Division laboratory. When word of the outbreak first came in, she was out of town at the annual meeting of the American Society of Microbiologists. As soon as she got back, she and the rest of the lab got straight to work. "The state epidemiologists had already looked for the routine pathogens, so we looked for those, but also for the unusual. We identified all the aerobic bacteria from the ills, and then, later, from the controls. We do everything very methodically," says Wells.[7]

Sometimes in this process something can appear that confuses the issue. They isolated *Klebseilla oxytoca* from some of the ills. "When we saw this in the stool samples from the patients, we got very excited," remembers Riley, who heard about the identification by phone. *Klebseilla oxytoca* is a bacterium often found in water where wood wastes from wood products mills have been deposited. That would make sense in Oregon.

Wells was not convinced. "What happens a lot of times, and what leads people astray," she says, "is that if someone has diarrhea and they isolate an organism and it's the predominant organism, they assume it's the pathogen. What happens when people have diarrhea is that the pathogen may not stay around very long, and there are organisms that are just good colonizers. When we did a further analysis of the *Klebseilla* in the ills, we found they were different strains." That would mean they came from different sources, she explained. "We also isolated *Klebseilla* from the controls," Wells says. At that point they lost interest in it.

The CDC had only six specimens from the Oregon patients. But in three of them they isolated something else—a rare serotype of *Escherichia coli*. It had the unwieldy name of *E. coli* O157:H7. Wells suspected at once that it might be the culprit. Her colleagues didn't agree, but she held out and just kept pursuing the unusual *E. coli* serotype. "We went back in our culture collection, and we found an O157:H7. It was from California. It had been isolated in 1975 from a patient with bloody diarrhea." That was a critical discovery. "This was a very unusual organism," Wells says. "You would not expect to find it in three out of six people. And we didn't find it in the controls," she explains.

The California patient from whom *E. coli* O157:H7 had been isolated had been ill with hemolytic uremic syndrome, but the significance of that wasn't obvious at the time. Joy Wells just stuck with what she was doing, a methodical retracing of a path worn smooth by the countless microbiologists who had come before her. Her strength was her refusal to be led off in the wrong direction. "We felt O157 might be the agent, but we didn't feel completely comfortable saying it was," says Wells. "We had no mechanism."

A "mechanism" is the method by which something causes a disease. Scientists don't feel confident saying something is responsible for an illness until they know how. At this point there was only an association between the outbreak of a disease characterized by bloody diarrhea and the rare serotype of *E. coli*. Often *E. coli* causes no trouble at all; sometimes it can even be helpful. The serotypes of *E. coli* that cause illness are divided into three groups, according to how they operate: enteropathogenic, enteroinvasive, and enterotoxigenic. *E. coli* O157 did not fit into any of these categories. Some serotypes are responsible for traveler's diarrhea, but that is a mild disease compared to what was happening in Oregon. These people had been very, very ill.

Riley completed his information gathering and headed back to the CDC. One thing became abundantly clear when the data were analyzed: Eating at McDonald's was strongly associated with the illness. But the fast-food chain served only USDA-approved meat. An association with foodborne illness, however tenuous, can have disastrous consequences for an eating establishment, which may or may not be implicated in the final analysis. There was no conclusive proof at this point, and that proof might never be forthcoming. The outbreak might just go into that thankfully small file of those that never really get solved. The epidemiologists tried their best to keep it quiet.

But Peggyann Hutchinson, a thirty-eight-year veteran of the *Medford Mail Tribune*, says that the fast-food association was no secret in the city. "We didn't hide it here. Everybody in town knew."[8] If the residents of Medford knew, it was because word travels fast in a small town. The coverage of the outbreak in the *Mail Tribune* ended on March 23 without a mention of either the McDonald's association or O157:H7.

It might have ended like that if three months later Bryce, Marcy, and

Lance Turner of Traverse City, Michigan, hadn't been taken by their father to a fast-food restaurant one evening in June 1982 when their mother was working. Almost a week later the children began getting sick. "The kids had really bad stomachaches and really bad diarrhea," their mother Lynn says now. Bryce, then eleven, was the worst off, and he was getting sicker. "His stools were pure blood." They took him to the hospital, where he was admitted. "I just remember him being hospitalized and IVs going and it was just very, very serious," Lynn Turner says now, trying to recall more than a decade later just what happened.[9]

At the hospital the doctors asked questions but didn't seem that surprised. They'd seen quite a few cases like Bryce's in the past few days. But they were certainly worried: These were very sick people. The doctors had strong suspicions—the link to McDonald's seemed to pop up often in answer to their questions—but they could find nothing in the lab test to explain the disease. Clearly they were in the midst of an outbreak of some kind. Clearly they needed help.

Dr. Robert Remis was then an EIS officer working with the Michigan Department of Public Health, and he remembers getting a call from the Tri-City Health Department in Traverse City asking for assistance. He and his boss, Dr. Harry McGee, knew this could be an important outbreak. Remis thought at once of the Oregon outbreak of bloody diarrhea he'd heard about through the CDC network. To complete the link to Oregon, the Tri-City Health Department had already made the association with McDonald's.

Remis had little doubt that his hunch was correct. "There is a dictum in our trade," he says. "If you go into an outbreak as an outside investigator and they don't tell you within the first fifteen minutes what the cause is, you'll never find it. Basically, what we really do is confirm what people already know."[10]

Before leaving for Traverse City, he called Riley. In the intervening months the CDC had been studying *E. coli* O157 further. It looked interesting enough to justify tests on animal models. Collaborating with the CDC veterinarian Morris Potter, the CDC investigators began with Rhesus monkeys, rabbits, and infant mice. They found nothing at first: The bacteria produced none of the standard responses they might have expected. That was discouraging. It's almost a tenet of modern medicine that what causes an illness in humans will cause an illness in animals. In fact, preventing illness in animals has long been a public health strategy in preventing illness in humans. Finally, in a test on infant rabbits, the bacteria did cause diarrhea—but not bloody diarrhea. All they could do at that point was watch and wait. If this unusual bacterium had caused the illness, it was unlikely to be an isolated occurrence. It would strike again.

In the labs Joy Wells was told about the Michigan illnesses, and at once she, too, suspected the outbreaks were related. "I was excited when I found O157," she remembers, "but when we had the second outbreak, I thought, 'This is it! Now we can prove it for sure.'"

They had the Oregon data, and the association with McDonald's had been strong. During the two weeks before the first sign of illness, the data demonstrated, twenty-one of twenty-five cases (84 percent) had eaten at one of the two McDonald's restaurants in town. Only thirteen of the forty-seven controls (28 percent) had eaten at the same restaurant. Three of the four ills who did not recall eating at that McDonald's remembered eating at another McDonald's within a week before the onset of the illness. And of those twenty-four who had eaten at McDonald's, twenty-one remembered eating the Big Mac.[11] Any of the well-known ingredients that almost any American could recite on request—"two all-beef patties, special sauce, lettuce, cheese, pickles, onions on a sesame seed bun"—might have been the culprit. "In the early days of an investigation," Wells says, "before you know anything, you just suspect everything."

In Traverse City, Bryce Turner had eaten a Big Mac. His brother and sister had eaten regular burgers. At first the CDC suspected the onions, with good reason. The onions were reconstituted from dry chopped onions and served without recooking. If they had been contaminated, microbes might have multiplied in the process. Remis had another idea. He thought it was the hamburger. But he remembers Dr. Mitchell Cohen, then assistant chief of the Enteric Diseases Branch, saying that it couldn't be. The meat was cooked. The McDonald's grills were kept at a temperature of 350 degrees, and there was no indication that the bacteria they were looking for were not susceptible to heat.

But Remis, a physics major in college, found that a thought kept nagging at him. McDonald's was careful with its meat patties. To make sure they were as fresh and safe as possible, McDonald's instructed its restaurant employees to keep the patties frozen in a cool cupboard near the grill; they were then tossed on the hot metallic surface with a glacial clunk. Remis wondered what a load of frozen meat patties would do to the grill temperature. He made some calculations, but there was only one thing to do to find out for sure. He and McGee returned to Traverse City with a primitive meat thermometer, timing their unannounced arrival to coincide with the busiest time of the week, Friday evening. The cooperative restaurant manager had a better thermometer that took instant readings from the grill surface. Remis's preliminary calculations had been correct. McGee took readings at sixteen reference points every minute and called them out to Remis. There it was. The proof. The cook was using part of the grill continuously—and it was cooling down. The burgers were not being thoroughly cooked. Remis felt vindicated.

But there was something else he had to do. He wanted to get a look at McDonald's meat-processing operation. The CDC decided to let both Riley and Remis make the trip. They met in Detroit and visited the Ohio plant together. It was a clean, efficient operation, Riley says, "so we really didn't think the contamination occurred at either the restaurant or the meat-processing plant."

Unless, thought Remis, the meat had arrived contaminated before being made into patties at the plant. The lot of meat consumed during the outbreak was long gone by this time, but the lot number had been recorded. While they were in the plant, Riley and Remis learned that the company had kept samples of the lot for quality control purposes. It was an incredible break. The samples were sent to the CDC labs.

Back in Atlanta tests had been run on the patient specimens from the Traverse City outbreak. The lab tested for the new *E. coli* serotype and found it—in about half of the samples. " 'This has got to be the agent,' " Wells remembers saying. "We felt that very strongly, even though we didn't have the mechanism." Then the lab tested the sample from the burger lot and Wells's team found what they had been looking for. They isolated *E. coli* O157:H7. Wells had to be sure. "We did some plasmid profiles, and the strains from Michigan, Oregon, and the hamburger meat were all the same."

They had discovered a new pathogen.

To describe a pathogen as "new" can mean any one of several things. It can mean that it is simply "new to us," new to our experience in a particular place and time. "New" can also mean that a preexisting microorganism that has never caused disease before suddenly finds, because of some change in the environment, a friendly niche within which to expand. Or "new" for a pathogen can mean precisely that: a completely new organism created either naturally through bacterial mutation or conjugation, by which microbes exchange genetic information, or through science, which, in the past twenty-five years, has developed the ability to create new organisms using techniques of genetic manipulation or bioengineering—taking genetic material from one organism and transferring it to another. *E. coli* O157:H7 could have been new in any of these three ways, but in the summer of 1982 Riley, Remis, Wells, and the rest of the CDC team could be certain of only one thing: It was new to them, and new to the medical literature.

The public wasn't to know. The first they heard about the new pathogen came months later when Drs. Riley, Blake, Remis, and Cohen presented the CDC's findings at an October medical meeting in Miami. Although they referred to McDonald's only as "Restaurant A," Steve Sternberg of the *Miami Herald* discovered from someone in the audience what that referred to. On October 7, 1982, the *Herald* ran a story headlined "Undercooked Burgers Linked to Disease." Sternberg quoted Dr. Mitchell Cohen: "Contaminated meat carrying previously unknown bacteria apparently caused the intestinal ailment called hemorrhagic colitis."[12]

The CDC tried to lessen the blow to the fast-food giant. Cohen said he suspected that hamburgers sold by other restaurants carried the bacteria as well and that any undercooked hamburger meat, including that purchased at a grocery store and cooked at home, might cause illness if the bacteria was present.

McDonald's put the best spin it could on the leak, downplaying the seriousness of the disease, the numbers of ill people, and the conclusive microbial link. Its spokesman, Raymond Caruso, told the *Herald*, "Some months ago the health officials at the CDC made us aware of the possibility of a statistical association between a small number of diarrhea cases in two small towns and our restaurants."[13] He didn't mention the identification of the pathogen in the lot of hamburger meat. Behind the scenes they were planning what they would tell the rest of the world. They knew the story was out. The wires picked it up at once, and the other papers would have it the next day, but the stock market had reacted immediately. Trading of McDonald's Corporation stock on the New York Stock Exchange was halted at 2:45 P.M. after it fell one and a half points. The next morning the fast-food giant made the headlines in the *New York Times:* "McDonald's Linked to 47 Stomach Disorder Cases."[14] Other New York papers were not so restrained: "A Big Mac Attack?: Hamburgers Linked to Intestinal Disease," the *Daily News* screamed.[15]

If stockholders were put off momentarily, they recovered. McDonald's did, too. "McDonald's Gets Break on Ailment," the *News* said, in an almost conciliatory headline the next day.[16] The stock had rebounded, climbing four and a quarter points in heavy trading. Nobody was going to mess with American burgers and get away with it. Investors leapt to the chain's defense with enthusiasm. A spot-check by the *News* "found that New Yorkers' appetites for McDonald's products seemed to be unabated."[15] But Cohen had been right. There was absolutely no reason to think that McDonald's was exceptional in any way. The CDC knew that O157 would strike again.

And because even small-town papers get the wire services, the people of Medford, Oregon, and Traverse City, Michigan, would finally discover what had hit them in the spring and early summer. "Illness victims ate at McDonald's," Allen Hallmark wrote in the *Medford Mail Tribune* on October 9. The local franchise owner called that information "ludicrous."[17]

As soon as O157 had been established as the culprit, the CDC looked around for other cases. Riley and Remis asked state and territorial epidemiologists to let the CDC know when they came across cases meeting the same criteria—crampy abdominal pain, grossly bloody diarrhea, absent or low-grade fever, and stool cultures negative for *Salmonella, Shigella, Campylobacter,* ova, and parasites. To get the word out further, they announced in various medical journals that they were looking for cases. "Then we started getting calls from various parts of the country," says Riley.

The CDC's *Morbidity and Morality Weekly Report* soon reported that hemorrhagic colitis caused by *E. coli* O157 was occurring sporadically across the United States. "We learned 90 percent of what we know today from those two epidemics," Riley said in 1993. When the outbreaks occurred, he says, "both establishments were offering promotions. I don't think there was any obvious breakdown in the food handling, but it's possible that there were a lot of hamburgers being cooked in those periods, and it's possible some of them remained insufficiently cooked."

"Once we identified meat, cattle had to be involved," says Riley. But they did not know for sure how the bacteria got into the meat. McDonald's tried to trace the beef back to its source, with no success. "It's not easy. It's a very complex network," says Riley. "The slaughterhouses, the distribution of meat items, it's practically impossible."

But the two outbreaks, the new pathogen, and the potential for further outbreaks of a serious hemorrhagic diarrheal disease set off a flurry of research in the medical and scientific communities. There was much to be learned, and everyone wanted to be the first to come up with the facts.

## Northern Exposure

The CDC wasn't the only agency to notice the new *E. coli* serotype. In Canada at the National Enteric Reference Center at the Laboratory Center for Disease Control (LCDC), a group of researchers had long been accumulating an interesting collection of what they called verotoxin-producing *E. coli*. The term was created by Dr. J. Knowalchuk, a Canadian food microbiologist who had reported in 1977 that the toxin produced by certain strains of *E. coli* killed a line of African green monkey kidney cells called vero cells. Since something that kills cells is called a cytotoxin, he called it verotoxin, or VTEC. It was quite distinct from the other toxins known to be produced by *E. coli*, but at the time it was a toxin looking for a disease. The Canadians knew the illnesses had to be out there, and they were looking for and collecting examples of the toxin-producing *E. coli*. They had already spotted the serotype O157.

Hemolytic uremic syndrome was first identified in 1955 by a Swiss physician, Dr. C. Gasser. It was characterized, Gasser wrote in the medical journal *Schweiz med Wochenschr*, by severe anemia, a marked decrease in the number of platelets in the blood, and the failure of the renal system.[18] Patients—usually but not always children—became very ill indeed. The illness followed, more often than not, a respiratory infection or diarrhea, sometimes bloody, but the actual cause of the syndrome was not known. Other physicians began noticing the syndrome, and any number of different agents, including drugs, toxins, chemicals, and microbes, were considered as possible causes. It seemed likely by the late 1970s that an infectious agent was responsible because cases tended to cluster in communities and families; an assortment of viruses and microbes had been proposed, but none seemed likely enough even to be put to study. The disease was a mystery. Between the late 1950s and 1960s HUS seemed to be on the increase, but it was hard to be sure. Often when something new has been identified, it is difficult to tell whether an apparent increase is real or simply the result of a new awareness. Still, a clinic in the Netherlands, for instance, had reported only four cases between 1959 and 1964 while between 1965 and 1970 there were fifty. Something seemed to be happening.

In September 1980 Dr. Brian T. Steele, then a thirty-five-year-old pediatric nephrologist at the Hospital for Sick Children in Toronto, had been on duty when several children were admitted with similar symptoms. He diagnosed them correctly as having hemolytic uremic syndrome, but he was surprised to see so many. "We were used to seeing four or five kids a year come in with HUS," he says.[19] All of a sudden he had five at once. And they would keep coming in.

Talking to the mothers as he worked them up, Steele quickly discovered a common link. Eventually he would have thirteen patients with HUS, and most of them had attended a local fair. They had also drunk apple cider. Gradually a story emerged. A vendor at the market had offered samples of free, nonfermented apple juice manufactured the previous day from apples that had been stored at a local farm. All of the ill children had tried the juice at the market. All but one, that is. That child's father worked at the farm and had taken a bottle of the cider home.

Steele says,

> We asked the father what the apples had been like, and he said, "I'm ashamed to say, but they were absolutely disgusting." The fair took place in September, and the apples are still green on the trees then, so they had used windfalls from the previous year. I remember saying to the guy, "What do you mean by a bit disgusting? I mean, were there brown areas in these apples?" And he said, "Look, Doc, there were more brown areas in those apples than there were green areas. They were absolutely rotten."
>
> In fact, the more you hear about apple juice, the more disgusted you'll get. They can use pretty shoddy apples. The other interesting thing was, the weekend before they made the batch, they had slaughtered a cow in the same barn. At the time we thought, "Well, that's interesting." What could the cow be carrying that would get into the apples and the apple juice? We had no idea what we were looking for.

Still, while the connection with the apple cider was clear—enough to be certain a pathogen was involved—they had no idea what had caused the HUS cases. The outbreak had caused a stir in the Canadian research community. At labs and facilities throughout the country studies were under way, variously looking for a fungal, chemical, viral, or bacterial cause for the illnesses. And more cases were being discovered. Two children had died of HUS in the past month in Alberta, and another was critically ill, but the cases didn't seem to be related to those in Toronto. Later, much later, the Alberta cases would begin to make sense. It is a cattle-raising area.

Then something serendipitous happened. Months later Steele was giving a presentation on the outbreak to a group of physicians, one of whom was a pathologist who had been abroad for a while and missed the widespread publicity the apple juice incident had garnered. He came forward and told Steele of an autopsy he had done on a child who died of HUS at about that time, and he remembered the father mentioning apple juice.

The child had not made it to the hospital. When his parents took him to the doctor because of his bloody diarrhea, they told the doctor that they had read in the paper and heard on television about the kids with kidney disease, and that their child had drunk the juice. The doctor responded with those immortal words, "You don't want to believe everything you read in the paper." The child was found dead in his bed the next morning. But even if he had been hospitalized, Steele was certain he could not have been saved. His organs had been destroyed. "That child," says Steele, "had the most florid hemolytic uremic syndrome you can imagine. His guts were infarcted—gangrenous. His kidneys were gangrenous as well; his brain was gangrenous. The thrombosis, or clots that form, can involve many organs. It was throughout his body, including his brain."

"The pathologist," remembers Steele, "had done a good job. He'd swabbed everything, and the swab had come back almost pure *E. coli.*" It had been identified as pathogenic and sent on to the labs in Ottawa, as all pathogenic strains were, but it had been misidentified. The *E. coli* group at the National Enteric Reference Center typed it as *E. coli* O113 verotoxin-producing. Research such as Knowalchuk's can lie buried in the medical literature, familiar only to a tiny group of highly specialized researchers. It may take years or even decades to trickle down to clinicians such as Steele. His reaction was, "What the hell is verotoxin?"

Dr. Mohamed A. Karmali, the chief microbiologist at the hospital, was very interested in Knowalchuk's VTEC-producing *E. coli* and what it might be doing. A few VTEC isolates had been identified from patients who'd had bloody diarrhea, and after the outbreak he asked Steele and other nephrologists to help him look for patients with this bacterium. In the next five patients they had no luck. Then they began looking for the toxin in the frozen fecal samples of the children who'd been ill with HUS. "Lo and behold," says Karmali, "we found it. The stools were actually loaded with the toxin."[20]

They had missed it before because everyone's stools are filled with *E. coli* and by the time the patients became ill the toxin-producing strains had done their work and weren't present in sufficient numbers to be obvious in a stool culture. When a culture is full of various *E. coli* growths, they may be difficult to isolate. But when Karmali knew what to look for, he began to find them by the painstaking method of picking out individual growths and culturing them. They found a number of verotoxin-producing *E. coli*, and in two of the fifteen HUS patients the *E. coli* was O157:H7. It was also found in two of the siblings of HUS patients who'd had diarrhea at around the same time.

Shortly thereafter, the Canadians would announce the link between O157 and HUS.

The Canadians and the Americans had come at the culprit from different angles, but they were on the verge of a similar conclusion. The bits and pieces of the puzzle were coming together. The Canadians felt strongly that this was their verotoxin; the Americans that O157-related hemorrhagic colitis was their disease; the Americans even had a different name for the toxin, which they called Shiga-like toxin (SLT), because it seemed virtually identical to that produced by *Shigella dysenteria* type I. (U.S. researchers have since concluded that it is in fact identical to the *Shigella* toxin, and from here on I will refer to it as Shiga toxin to avoid the confusion of the dueling names.)

Pride, ambition, and chauvinism were playing a role, perhaps even contributing to the process. The researchers were nudging each other forward, whatever the motivation, by whatever means, enlarging in increments the pool of information about Shiga toxin, *E. coli* O157:H7, hemorrhagic colitis, and HUS. Gradually, in the awkward, haphazard way science is done, the story of a new pathogen was being written.

## An Emerging Health Threat

At the CDC the Enteric Diseases Branch watched and waited. The new pathogen was out there; it was simply a question of where and when it would next strike. It happened on September 16, 1984, in Papillion, Nebraska. Patients at a nursing home began getting ill with severe, often bloody diarrhea. Because older patients are vulnerable, the staff contacted the Douglas County Health Department; Sarpy County, where the home was located, didn't have a health department of its own. The Douglas County Health Department realized it needed help and turned to the state. Gary Hosek, a registered sanitarian with a master's degree in public health, was asked to investigate. Tests for *Salmonella, Shigella, Campylobacter,* ova, and parasites were negative. The department preferred to handle the outbreak on its own, but when the number of patients reached twenty-two, with half of them hospitalized, it called the CDC.

Robert Tauxe was put in charge and began coordinating directly with the county health department. He had his suspicions as to what it might be. He arranged for the stool samples to be shipped to the CDC, and to make certain of the results, the samples were given to two different labs. Both labs isolated O157:H7 in at least one sample.

Tauxe felt that surge of excitement common to epidemiologists. This was the new bacterium they had been watching for. But in the time taken to process the cultures, two patients had died. That hadn't happened before. Even though the patients were more vulnerable because of their age and other health concerns, this looked very serious.

When you don't know what is causing an illness, you don't know whether it can spread. There is always the possibility that even a few related cases are signaling the start of an outbreak. The guilty food could be in the community—something widely distributed and widely available. Quick action could prevent a further outbreak of disease. Tauxe and Dr. Caroline Ryan, a new EIS officer, were assigned. They loaded up the Igloo coolers that went on every outbreak and set out at once.

On the way out Tauxe went over what they knew and did not know about the bacterium. (He still has his list of questions.) It had been found in meat once, but there was, as yet, no reason to think that meat alone posed a danger. With so little experience with the organism, even what they knew was suspect. What might be the case in one outbreak might not apply to the next. What they would have to find out was how widespread the outbreak was. Did it extend to the larger community? Was there a common exposure? What were the clinical signs? Just how sick were these people? How long had the patients continued to shed the bacteria? Was any treatment having an effect? What was killing the patients? And if the source were an animal product, could it be traced? It was pointless, even counterproductive, to speculate on the answers to these questions.

There was an eerie moment after they arrived, Ryan remembers. They had just driven up to the motel. Looking off in the distance, Ryan saw a water tower with the town's name and, just beneath it, the golden arches of McDonald's. It gave her a start. McDonald's would get a break: No one at the nursing home had eaten fast food. The illness was confined to the nursing home. The investigative challenge was working with elderly residents who couldn't remember what they had eaten. One lead: The nursing home got its eggs from a nonlicensed local farmer. But tests on the farm were negative. Finally Ryan had the idea of looking at special diets to see whether there was a relationship to those patients who had been sick. There was. Of the nineteen in whom the disease was confirmed, seventeen had been on the low-salt diet. And on September 13, when the residents on the regular diet were having ham, those on the low-salt diet had eaten hamburgers. That wasn't proof, but tracking the hamburger told them more. It had come from Great Plains Meat Distributors in Nebraska City, Nebraska. Running short on meat from its usual suppliers, Great Plains had purchased 500 pounds from Nebraska City Wholesale Meats, then sold 40 pounds of that supply to the nursing home. Of the 200 pounds the distributor sold the public school system, 130 had been returned because of poor quality.[21]

Tauxe and Ryan had checked area hospitals for patients with similar symptoms, then set up an informal surveillance system at six hospitals. And they'd gone back through emergency room records. Several patients met the criteria they were looking for. One was a thirty-four-year-old mother from Auburn, Nebraska. Her family had cooked hamburgers around September 13, and she had subsequently become ill with cramping and loose stools, turning to bloody diarrhea, on the eighteenth. She was treated with antibiotics and after a few days seemed to improve. Then she developed HUS and died eleven days after first being taken ill. The hamburger she ate had been purchased in bulk from Nebraska City Wholesale Meats as well. Ryan wondered why other members of the family, who'd eaten hamburgers prepared from the same meat, hadn't become ill. The answer was simple: they liked their burgers thin, the victim liked hers thick. Her hamburger hadn't been cooked all the way through.[22]

In their final report to the Nebraska State Department of Health, Ryan and Tauxe concluded that the outbreak was caused by *E. coli* O157:H7 and that, once again, the most likely source was hamburger meat. It could not be proven. Not surprisingly, tests were negative for the bacteria on meat samples. The bacteria were elusive. The evidence was epidemiological, and the Auburn woman was the vital link. In their final memo Tauxe and Ryan urged continued surveillance, reporting of suspect cases, and, if possible, identification of the herds of cattle from which the suspect meat came. They recommended that both the cattle and their feed be cultured. What the USDA actually did, if anything, will never be known because in the 1980s, under new paperwork reduction regulations, documents were no longer kept more than two years; the record was destroyed.

It was the early 1980s. The emergence of AIDS had changed the way medicine was reported. Medical coverage had gone from the science page to the front page. AIDS would also strike at our complacency as if specially ordered for the purpose. It would signal to us that our apparent mastery of disease was an illusion, and that factors outside our influence might be at work in the environment. It would call into question our basic assumptions about a basic human act: sex. The emerging foodborne pathogens would call into question another basic and essential human behavior: eating.

*E. coli* O157:H7, says Tauxe, had something else in common with AIDS: It "challenged the central dogma of modern medicine."[23] It masqueraded as other diseases, was often misdiagnosed, and produced little or no fever. While O157 and HUS had apparently been around for some time, no one had perceived the illnesses as infectious. Most agents that cause disease in humans also cause it in animals. This was not true of O157. And finally, says Tauxe, it upset the kingpin of the central dogma: Antibiotics fix infections.

"The only problem has been when bugs become insensitive to antibiotics, and yet when antibiotics were given for *E. coli* O157, it was very difficult to see any benefit, and some have concluded it may cause harm," he says. It wasn't that the bacteria were resistant; it was that antibiotics had no effect on the toxins the bacteria produced. And it was the toxin that was damaging cells and making people so ill. Indeed, even seemingly harmless antidiarrheal medicine appeared to make things worse. It seemed that diarrhea had a purpose: ridding the body of something harmful. Antidiarrheal medicine prevented that.

At nearly the same time as the Papillion outbreak, another was taking place in a day-care center in North Carolina. Thirty-six children out of 101 had diarrhea, and *E. coli* O157 was identified as the cause. Three children were hospitalized for HUS, but the disease was relatively mild. Hamburger was not implicated in the outbreak, No one knew where the first case had come from, but it seemed clear that it was transmitted from child to child. That worried the researchers. If this bacteria could be spread so easily by person-to-person contact, it meant that the number of organisms needed to make someone ill was very small indeed. The lower the infectious dose, the more contagious the disease. Nursery schools and day-care centers might be particularly vulnerable. In a changing culture in which day-care centers have become more and more important to the functioning of the economy, this had serious implications. There would be further day-care and kindergarten outbreaks. In one, the wall-to-wall carpet was implicated as the route of cross-contamination. The health implications of having so many children on a surface that could never be adequately cleaned had never been seriously questioned before.

There is a worldwide network among epidemiologists. Word travels fast even when material has not been published. There had been virtually nothing about the new pathogen in the national press, but even in the mid-1980s the CDC knew it was spreading. It was hearing of outbreaks in Canada, where the LCDC was keeping close tabs on the new pathogen. According to the *Canadian Disease Weekly Report*,[24] eighteen patients at Calgary General Hospital in Alberta had the now-familiar symptoms. All were positive for the pathogen. Then Dr. Anne Carter of the Bureau of Communicable Disease Epidemiology at the LCDC investigated a nursing home outbreak in which nineteen of fifty-five patents with *E. coli* O157 infection died; moreover, the death rate among those in whom the disease had progressed to HUS was a truly frightening 88 percent. Researchers noticed that patients taken antibiotics seemed to be more susceptible to person-to-person transfer in the second wave of the outbreak; they were also more likely to die. Carter believed that gastric activity had a role in the body's defenses against *E. coli* O157:H7 and that something that disturbed that activity, such as antibiotics, could make people more susceptible. Could antacids work in the same way? she wondered.

Looking back from what we know now, it's difficult to imagine that stage when what was known about the pathogen was so limited. But as outbreaks were investigated, each seemed to reveal something new. The CDC had attempted to relay some information to the public in *MMWR* reports, but the key was getting the press to pass it on. Journalists didn't seem very interested in writing about diarrhea, however, perhaps because it is so common and usually not terribly serious. But the outbreaks didn't stop.

In 1986 three older women from Walla Walla, Washington, a town in the eastern part of the state, were transferred to a Seattle hospital. They had thrombotic thrombocytopenic purpura (TTP), an adult version of HUS. An alert physician notified Stephen Ostroff, an EIS officer recently assigned to Washington State; after notifying the CDC, Ostroff set off to investigate.[25]

He would discover that a dozen people had reported to clincs and emergency rooms with bloody diarrhea in the weeks before, but the medical community in the town of twenty-five thousand had neither noticed the increase nor made the link between them. Clearly something was going on, and it looked like another outbreak from *E. coli* O157:H7. Another EIS officer, Dr. Patricia M. Griffin, was dispatched to help Ostroff with the investigation. The interviews implicated a favorite Walla Walla Mexican-style restaurant called Taco Time. It was a complicated investigation because so many foods contained similar ingredients: cheese, tomatoes, lettuce, and ground meat. When a nursing home nearby reported diarrhea cases and both the home and the restaurant were found to have used meat from the same source, the puzzle was solved.

One thing had been helpful since the beginning of this outbreak. The strain was unusual. Ostroff and Griffin would later discover that an HMO in Seattle had been doing a survey of the causes of diarrhea at around the same time, culturing every specimen, and had found a huge peak in the isolation of the same strain of *E. coli* O157:H7. There has also been children ill with HUS in Children's Hospital in Seattle. Ostroff and Griffin would realize that the contaminated meat had been distributed throughout the state and had actually caused a statewide outbreak that health officials had completely missed after the fact.

*E. coli* O157:H7 was one of the first things Griffin heard about when she joined the EIS. The Walla Walla outbreak was her first chance to see what the bug did in real life. and what she saw was frightening. When she and Ostroff discovered that hamburger was the likely cause, they notified the USDA to attempt a traceback. It was no easy task. Hamburger in the supermarket may appear to be meat from one source. It seldom, if ever, is. Most supermarkets buy coarse ground meat in ten-pound chubs and regrind it, sometimes adding scraps of their own. Distributed by a processor, the chub may represent cows from many farms and even different countries. It is prepared in forty-thousand-pound lots or greater. Three possible sources of the meat were implicated by the Walla Walla investigation. Two were large producers; the third was a small dairy processor in the southwestern part of the state.

Most people prefer not to think about what happens to dairy cows that no longer produce enough milk to please their owners. They usually became hamburger meat. Some slaughtering facilities specialize in these animals. After days of going through tedious records, Griffin came up with the most likely farms, and in a matter of days she and Ostroff would be slogging through mud swabbing cows with the help of EIS officer Larry Shipman, who was also a USDA veterinarian. They would find the organism in only about 1 percent of the 539 animals they tested. If hamburger were produced from a single animal, the chances of its being contaminated were apparently small. Mass-production and distribution had made a huge problem out of what might have been a serious but small one.

The investigation complete, Ostroff and Shipman would go on to other things. But Griffin found herself hooked on the new pathogen. It would become her bug. The growing number of outbreaks were of grave concern in the Enteric Diseases Branch. Griffin realized that her training had been in gastroenterology, but she'd never heard of the organism before coming to the CDC—nor, she discovered, had her former colleagues at the University of

Pennsylvania and Harvard's Brigham and Women's Hospital. And yet *E. coli* O157:H7 was apparently an important cause of bloody diarrhea—something she had seen. How many cases had been misdiagnosed? The idea that no one in the field of gastroenterology seemed to be aware of this important pathogen was not just astonishing, it was intensely frustrating. The link to HUS that Mohamed Karmali had made in 1983 made sense. Autopsies of HUS victims showed the cellular destruction in the tissue that seemed to be caused by toxin. Few in the Enteric Diseases Branch had any doubt of the connection. Griffin felt strongly enough to start making calls around the country to tell pediatric nephrologists and to propose a study to look for the organism in the stools of HUS patients. She met resistance. On the whole, the doctors were skeptical and indifferent—understandably so, says Griffin charitably. The syndrome had been identified in 1955, and so many theories as to its cause had been advanced that this seemed like just one more. "None of them thought we knew the cause, and only a few thought it was worth going on another wild goose chase," says Griffin.[26]

Then Dr. Marguerite Neill, then an EIS officer in Washington State, and Dr. Phillip Tarr in Seattle did a study that showed that most cases of HUS were related to *E. coli* O157:H7. It was the evidence the CDC needed to make an even stronger case to physicians. Griffin had her mission.

Six years later, when the Jack-in-the-Box outbreak occurred, Griffin would be the expert the world turned to. Before long the network of people whose lives would be touched in some way by this pathogen would be calling her the "High Priestess of *E. coli*."

## Cherchez la Vache

Between 1986 and 1992 the public remained virtually unaware of the new pathogen except when it struck in their own backyard, so to speak. And then the information about outbreaks or sporadic cases remained local. To make it worse, perhaps in a misguided attempt to simplify or describe in familiar terms for the public the serious involvement of the kidneys in HUS, the syndrome was at first described as a urinary tract infection in some news reports and even, on occasion, by the USDA, giving little hint of its horror.[27] Behind the scenes in research circles, however, there was much activity. When CDC investigators looked further for the pathogen in herds, they found it. The accumulating evidence pointed to a bovine host and the pathogen could be carried on products produced from cows—either meat or milk. It got into these products through contamination from the animal's fecal waste.

Despite its genuinely nasty traits, *E. coli* O157:H7 did have one desirable quality: It did not ferment sorbitol. That meant that a culture medium containing sorbitol could be prepared in which it would show up easily. Sorbitol-MacConkey quickly became the standard culture medium, but few labs were using it yet. When they did, they discovered that the pathogen was by no means rare as a cause of human illness. A study in Washington State showed that the rate of isolation was comparable to that of *Campylobacter, Salmonella,* and *Shigella*. The state would soon make O157 a reportable illness.

In the meantime, other researchers wanted to know where the new pathogen had come from. They were looking at its genetic makeup. What they found was odd. The one hundred strains of *E. coli* O157:H7 were quite different genetically from the other Shiga toxin–producing *E. coli,* yet they weren't very different from each other; they didn't show the wide diversity that would have developed over a long period of time from normal mutations and adaptations. What could be concluded from this was that they seemed likely to represent a single clone that had recently descended from a common ancestor and become widely disseminated geographically.

What had distributed it? Was the vast, worldwide trade in meat a possibility? There are reference centers where samples of bacteria are stored. An O157 that had been cultured from an Argentinian calf during the 1970s was found among them. Argentina was both an exporter of beef and a country with a high incidence of HUS. Could it have begun there? But now wherever scientists from North and South America, Europe, and Asia looked for it, they found it. There were places in Canada where it was isolated more frequently from stools than *Shigella*. It had been isolated from Korean and Chilean children and from Thai adults. It had been found in England, the Netherlands, and Germany. It was new, it was different, it was virulent, and it was everywhere.

On the clinical side, studies were revealing that the bacteria didn't stay long in the human body. If you wanted to have a good chance of finding them, you had to culture patients within seven days of onset of illness—and before they had taken an antibiotic, as most strains were still not resistant. After that, it was shed, leaving the toxin behind to do its dreadful work.

Researchers concluded that when patients got HUS, it seemed likely that the toxins damaged the endothelial cells, causing damage to the intestines, which created the bleeding and also allowed the toxins to enter the bloodstream. Then arteries and capillaries became choked with the shredded particles of red blood cells and platelets destroyed by the toxins. In the kidneys the shredded material collected, disrupting their function. Body organs, because blood could no longer reach them through these clogged passageways, became starved for oxygen, swelled, and were damaged. Frustrated physicians had no treatment for this vicious, cascading pattern, only support—dialysis, respirators, IV fluids, transfusions—as patients' bodies fought the toxin, sometimes winning, but in 3–7 percent of cases failing before their eyes.

It was six years after the first O157 outbreak, and researchers were discovering more and more each year about the new pathogen. In 1988 an outbreak of *E. coli* O157:H7 infections in fifty junior high students was traced by the Minnesota team (there seemed to be something going on in the northern tier states) to precooked hamburger patties served in the school cafeteria. The patties appeared to be cooked enough to be served without further cooking, but they had not been cooked enough to destroy the pathogen.

With the encouragement of the Minnesota epidemiologists, the USDA at this point tried to raise the temperature to which the burgers had to be precooked; the agency ran into a balky industry. The meat industry feared new regulations and believed, probably correctly, that precooking a beef patty that was supposed to be recooked at home or school to the 160 degrees proposed by the USDA, and supported by the Minnesota epidemiologists, would leave it tasting like a hockey puck. It would also reduce the water content and thus the weight—the Holy Grail of profit-making in the food industry. The American Meat Institute resisted. Dr. George D. Wilson said the association saw no "grounds for associating a public health emergency with the production and consumption of precooked meat patties." The Western States Meat Association said that the proposed requirement of a minimum internal temperature for fully cooked patties would make it impossible to prepare a medium or medium-rare beef patty.[28] Little did they know that in a few years eating a rare beef patty would be considered the culinary equivalent of Russian roulette. The voice of industry is loud and powerful. The temperature to which patties had to be cooked before distribution wouldn't increase until 1993.

While hamburger appeared to be the chief villain for O157 infection, there were reports of outbreaks that seemed to have different causes. Several were linked to drinking unpasteurized milk, not unexpectedly, since cattle harbored the bacteria. People who had handled potatoes stored in peat (presumably contaminated with manure) were infected in England. In 1990 the largest outbreak to that date occurred in the small town of Cabool, Missouri, where the municipal water supply became contaminated after a hard freeze broke water pipes and allowed contamination. The outbreak, which took two months to control, made more than 243 individuals ill, hospitalized thirty-two, and killed four. Another deadly episode followed when two outbreaks in institutions for the mentally retarded in Utah in 1990 caused twenty illnesses that resulted in eight cases of HUS and four deaths.

Then, as if to confuse the issue, an outbreak in the late fall of 1991 in Massachusetts that resulted in four cases of HUS was found to be linked to drinking fresh-pressed apple cider. To members of the beef industry, who had been growing increasingly nervous, this was good news. They could point to apple cider and say, "See! It isn't just beef." But cows were probably ultimately responsible for the Massachusetts outbreak. The apples were likely to have become contaminated on the ground from cow manure—just as they apparently had in the 1980 outbreak that produced fifteen cases of HUS in Canada. Then an outbreak in Oregon from swimming in a fecally contaminated lake demonstrated that infection could be linked not just to a food but to an activity. Future investigations of outbreaks, noted Neill, would require creativity and insight. One wag had a word of advice: "Cherchez la vache." But it also seemed that the pathogen was now in the environment, not just in cows. It could pop up anywhere there was danger of contamination from fecal material, and that seemed to be almost everywhere.

In terms of outbreaks, 1992 proved to be a "relatively quiet" year for the pathogen, Dr. Marguerite Neill, now at the Brown University School of Medicine, told a group of sanitarians in 1994. In fact, sporadic cases were causing heartbreak. Mary Heersink certainly would never forget 1992. Nor would Robert Galler and his wife Laurie of Long Island, New York.

Their daughter, Lois Joy, had become ill on Thursday, June 26, 1992. By the next day she had a stomachache. That was followed on Saturday by frequent and severe diarrhea, which by Sunday had become bloody. The family's pediatrician saw the child and sent her to the emergency room because she was dehydrated. At the hospital they rehydrated Lois, did a blood workup, and took X-rays of her stomach. Then they discharged her. They would take her back on Tuesday, "more and more lethargic with each passing hour," remembers Bob Galler.[29] One of the first questions the doctor asked was whether she had recently been to a fast-food restaurant or eaten hamburgers.

Then commenced, Galler remembers "the most emotional roller-coaster ride that any parent can experience." His daughter's kidneys stopped functioning on July 2, "never to function again. During her eighteen days in the hospital," he says, "she was given sixteen blood transfusions, fourteen with fluid, she had to be put on a respirator, she lost sight in her right eye, she suffered an infarction to her brain. We watched, simply helpless, as our daughter died before our very eyes."

On Wednesday, September 8, 1992, a similar ordeal began for Arthur O'Connell, a high school mathematics teacher from Kearny, New Jersey, a town about twenty minutes outside of New York City. Katie, his twenty-three-month-old toddler, had diarrhea.[30] By the next day she had dry heaves and was lethargic throughout the day. When her diarrhea became bloody, the family took Katie to the pediatrician, who immediately hospitalized her. The hospital began a series of tests; on Thursday Katie went into a seizure. Tests revealed HUS, and she was transferred to another hospital where she was put on peritoneal dialysis.

"On Tuesday Katie had the tubes removed from her nose. She opened her eyes and talked to me," remembered O'Connell. "I asked her if she wanted anything, and she told me, 'Apple juice.' Those were the last words she ever said to me." Then, in the now-familiar pattern, crisis followed crisis until their daughter stopped breathing and was hooked to a respirator. She died on September 16, 1992. Scattered across the country—and around the world—were other deaths no less tragic and unnecessary.

In 1992 the food industry belatedly entered the picture by funding research on *E. coli* O157:H7 through the International Life Sciences Institute, which is supported by a wide range of corporate food giants. But lurking just over the horizon was the Jack-in-the-Box outbreak, which would finally galvanize the press, the public, the industry, and the regulatory agencies. Before the year was out Lauren Rudolph would go to a fast-food restaurant with her father, and the massive, widespread tragedy that would be called the western states outbreak would begin.

On January 12, 1993, a gastroenterologist at Children's Hospital and Medical Center in Seattle notified the Washington State Department of Health that physicians there were seeing many children with bloody diarrhea and some with HUS. Investigators quickly did a preliminary investigation and discovered almost at once that most of the cases, but none of the controls, had eaten at Jack-in-the-Box. Most had eaten a regular hamburger, which, being small, is a favorite with children.

The public first learned of the cases on January 17, when the *Seattle Times* and the *Seattle Post-Intelligencer* reported forty-five cases of foodborne illness in the western Washington area. The cases were linked to hamburgers. One day later the Washington Department of Health issued a public announcement urging people with bloody diarrhea to see a doctor. Hamburgers were recalled by the restaurant chain. The CDC would later estimate that the quick recall prevented eight hundred additional illnesses.

The beef industry was immediately on alert. Its PR people went into a huddle to plan what they should say and do. On January 19, Wendy Feik Pinkerton and C. J. Reynolds, spokespersons for the National Livestock and Meat Board, sent a memorandum to Washington State Beef Council executives telling them of the reports. "Here's what we know," they said.

> Of the reported cases, 13 children were infected and five of these are in intensive care. Of the reported cases, 37 people had been interviewed by the State Health Department; 27 had eaten at Jack in the Box restaurants in the Western Washington area and all 27 had eaten hamburgers. Jack in the Box has been the only restaurant chain implicated in this *E. coli* O157:H7 outbreak.

She also told the meat industry executives, although it made no sense at all and was clearly wishful thinking, that "Washington State epidemiologists said undercooking the burgers, and not the burgers themselves, appears to be the reason for the foodborne illnesses." Then she added: "While we do not anticipate widespread, national media attention, we have planned for it. Our goal, first and foremost, is to stay out of the media spotlight. The coverage, so far, had focused on cooking procedures at the fast food outlets, not beef industry issues. Let's try to keep it that way."[31] She had outlined the strategy that the industry would hold to. Hamburgers would be redefined.

Americans had been cooking and eating pink hamburgers for years; overnight, history would be rewritten. That would become "improper preparation." And the fault would rest with the consumer, who should have known better, although how the consumer would have discovered this mistake was mysterious and unclear. "There has been a subtle turning of this on to the consumer," says Steve Bjerklie, former editor of *Meat and Poultry* magazine, "and it's morally reprehensible."[32] Actually, it wasn't subtle at all.

It did not take Jack-in-the-Box's parent company, Foodmaker, long to trace the contaminated meat to a batch of seventy-seven thousand hamburger patties processed by Vons Company, a meat-processing plant in Los Angeles. Most had been shipped to Seattle, but ninety-two hundred were unaccounted for. They might have gone to San Diego, Los Angeles, Bakersfield, Las Vegas, or Hawaii; the company had no way of knowing. Two days later the *Seattle Post-Intelligencer* would report that while sixty-six people were already ill, the USDA had yet to begin tracking the source of the contaminated meat—for indeed, it was contaminated. The Vons spokesperson told the press that no changes had been made in the processing procedures. "The USDA maintains an office in our plant. There are not out-of-the-ordinary USDA tests going on."[33]

Both Philip Tarr, a pediatric gastroenterologist, and Dale Hancock, a Washington State University epidemiologist and veterinarian, were already experts on O157. They told the paper they were worried about the lack of testing. "If there is a common source of infection," Tarr told the paper, "then it should be pursued. If there is a delay, one would be concerned about further spread of the infecting pathogen." Hancock was even blunter. "You'd think that an organization with a name like 'Food Safety and Inspection Service,' would certainly monitor what happens at the plant."[34]

By the end of the month, three hundred were ill and it had become clear that the outbreak extended to four states. When the final count was complete, the case numbers would climb. Gaping holes in the public health system were revealed. Nevada, which actually had a reporting requirement for cases of O157, would not know it was in the midst of an outbreak until a parent of a child sick with HUS who had eaten at Jack-in-the-Box read the newspapers, made the connection and brought it to the attention of health officials.

The public was growing panicky. Hamburger sales were down. On January 21 in San Diego County, where Foodmaker has its corporate headquarters, the health department issued a calming release: "You can still eat at restaurants; you can still eat at home. Just be sure to have any meat cooked thoroughly." If the health department had made the link between Lauren Rudolph, who died from the pathogen in San Diego County three weeks before, and what was happening in Seattle, it didn't say so. The press would finally make the connection.

To the American public, the Jack-in-the-Box outbreak was about a new health threat from an ordinary, everyday food source, and it was about impressive numbers. (Although most reports stopped at three hundred and never updated their stories with the final figures.) The reaction, however, was not unlike the response after a hurricane or earthquake in some other part of the country: interest, concern, but ultimately the sense that "that was there and I am here." What the public did not understand was that the pathogen was everywhere and that everyone was a potential victim, for the routes of possible infection were now broad and varied; sporadic illness had become common across the United States. For those who were ill and those who lost children or suffered through weeks or months of hospitalization of their loved ones, it was about individuals, and it was pure hell.

Dorothy Dolan watched not one but both daughters, four-year-old Mary and three-year-old Aundrea, become terrifyingly ill after eating hamburgers from Jack-in-the-Box. Both children followed the familiar pattern of lethargy and fierce diarrhea, but Mary was the most severely ill, with HUS following its predictably horrifying course. Although the prognosis was poor, Mary, as well as Aundrea, would eventually recover. But after her long battle, Mary would need physical therapy, speech therapy, and occupational therapy. Six months later her blood pressure remained high and her speech was still slow. The Dolans would be left to wonder about the effect of the illness on her cognitive abilities, the risk of HIV from the transfusions, and her long-term kidney function. Even Aundrea, who fared better, would need routine testing.

Dorothy Dolan told her story to a Washington symposium in September 1993. It had been organized by Safe Tables Our Priority (STOP), a group of parents and friends of victims of *E. coli* O157. STOP would start a hotline for new victims, create a network, and act as a clearinghouse for information. With the help of the Safe Food Coalition, a group of organizations interested in food safety, it has become professional and organized, carrying its message to Washington.

"Not a day goes by," said Dolan, "that I don't hear about another child sick from *E. coli* and hemolytic uremic syndrome. I cry and feel sick to think another child and family has to go through what we had been through. We cannot continue to let our children suffer and die from something that can and should be prevented, from something that is stamped USDA-approved."[35]

The press, known for sensationalizing events, seemed strangely restrained to the point of disinformation. *Boston Globe* reporter Michael Rezendes on January 31, 1993, described Joseph and Dorothy Dolan as "the parents of two daughters laid low after eating cheeseburgers."

The Washington symposium was an emotional gathering. The Dolans could count themselves lucky. One child in the outbreak had been released from the hospital only to die in his car seat on the trip back to the hospital. Another family, Diana and Michael Nole, lost their only child, Michael James Nole. He was just over two when he died in January 1993 from a children's meal he'd eaten at Jack-in-the-Box. Like Dorothy Dolan, Diana Nole was a nurse. "Some of the things my son went through were the most horrific things I have ever seen in my eight years working in the medical field and my most recent two years working in an emergency room."[36]

Ten-year-old Brianne Kiner survived after spending forty days in a coma; her colon, kidneys, lungs, heart, pancreas, and liver were damaged. After five months in the hospital and a miraculous recovery, she was still, a year after the episode, suffering seizures and had been rehospitalized a number of times. She was still receiving special nutrients through a feeding tube in her stomach because all of her large intestine, damaged by the bacteria's toxin, had been removed, and several times a day she inhaled medication to keep her lungs functioning as another machine applied gentle pats to help loosen the mucus in her lungs. With her pancreas only partly functional, she endured several needle sticks a day to monitor her blood sugar levels and receive insulin. Yet she was there at the hearing.

Her presence, her few words, and the testimony of others would leave the audience damp-eyed and badly shaken. But they were mainly supporters and industry note-takers, along with a very few congressional aides. Despite the widespread attention the outbreak had garnered, few reporters or members of Congress bothered to attend. The public was fickle. The problem was behind them. But another mother at the hearing could have told them otherwise. Janis Sowerby of Saranac, Michigan, told the story of her son's illness following a camping trip with his father months after the Jack-in-the-Box outbreak. They had bought hamburger at a local store and prepared sloppy joes. His symptoms followed the same appalling, treacherous, downward spiral from diarrhea and cramps to bloody diarrhea, to HUS, to multiple complications and organ failure, to death.

STOP was disappointed by the small turnout for the symposium, but not discouraged. Its campaign would be waged at the USDA and with individual members of Congress, and much later, when President Clinton signed a bill to improve meat safety, he would credit the work of these parents, who had never given up the fight. But that would be three years away. In the meantime, the press and the public were beginning to ask what the USDA was going to do about meat safety.

## What Is Safe Meat?

If a piece of raw meat harbors bacteria that can cause illness and death, should it be considered adulterated or contaminated? The answer to that question has preoccupied a frightened meat industry, caused deep divisions in the U.S. Department of Agriculture, possibly contributed to the short tenure of the last secretary of agriculture, and ultimately exposed the distasteful politics of food. In the meantime, 81 million Americans—or 260 million, according to which expert from the Centers for Disease Control and Prevention one listens to—continue to be made ill each year by foodborne bacteria, and the number is increasing.

Agriculture Secretary Michael Espy had no sooner been sworn in than he was faced with an epidemic that would eventually swell to 732 illnesses, 195 hospitalizations, and 55 cases of associated hemolytic uremic syndrome. There were four deaths. The meat responsible had been inspected and passed by his agency. It was a galvanizing moment.

The impact this pathogen was bound to have on the food industry has been obvious from the beginning, both to the industry and to the USDA. The pathogen is harbored by healthy cattle. Why it is so difficult to find in cattle (although Dale Hancock, a researcher with Washington State University, says he has been able to find it in every herd he's looked at, in about 1–3 percent of the animals) but relatively easy to find in human patients with diarrhea has to do with a number of factors. When other researchers began looking at cattle, they found any number of places along the path from field to fork where the bacteria could be amplified. An infected cow might shed the pathogen in the field and the next grazing cow pick it up. Even if only one or two cows in a herd were shedding the pathogen at the time of shipping, tired, frightened, and hungry cattle (they are given neither food nor drink during transport) attempting to keep their balance in a moving vehicle lose control of their bowels. In these tight quarters the fecal material gets on the hides of other animals. When they arrive, all the cows are filthy and distressed. Conditions before slaughter favor the pathogen as well. Routinely cattle, massed together in tight quarters, standing in the excrement of many animals, are starved a day or so before slaughter to empty their intestines. Research has found that this practice also produces more shedding of the bacteria. Given the rapid pace of meat processing today (as many as three hundred animals an hour, or five per minute), it's not surprising that O157 and other pathogens get onto meat during the slaughtering process through contamination with fecal material. Combining meat from many different sources in ground beef can spread any pathogens present throughout the product. Robert Tauxe has described the hamburger of today as "a mixture of one hundred different cattle from four different countries."[37]

Up until a few years ago slaughtered meat imported into the country had to be a certain size—large enough to identify what it was. That size restriction was done away with in 1993, and now, in effect, we import hamburger or coarse ground meat. The idea that the tons of ground meat and scraps imported into this country can be precisely identified and examined for the presence of pathogens is pure fantasy. Much of our imported meat comes from Australia, where outbreaks from O157 and other toxin-producing *E. coli* have occurred recently although there is no reason to believe that Australian meat is any more contaminated than domestic meat. In fact, we export about as much meat as we import in a mindless global transfer of not just meat but infectious agents.

While the pathogen may be only on the surface of steaks and roasts, where it will most likely be cooked away (if the meat is not prodded to tenderize it or pierced with a fork during marinating or cooking), grinding spreads the bug throughout ground beef. Ten years of research have revealed no guaranteed way to assure the consumer of an absolutely safe hamburger other than thorough cooking, although even industry representatives concede that more careful skinning and slaughtering could help reduce the microbial load of beef.

The microbe *E. coli* O157:H7, prior to 1993, seemed to put both federal agencies and the meat industry into a state of deep denial—at least when it came to taking action. While the USDA as early as 1988 suggested to consumers who were curious enough to call and ask for information that hamburgers should be cooked to 160 degrees, the FDA allowed restaurants to follow a 140-degree guideline until the Jack-in-the-Box outbreak, when it rushed to raise it to 168.3 degrees in the model food code, which states adhere to either voluntarily or, in some cases, not at all. (Temperature tests of hamburgers at one Jack-in-the-Box restaurant taken after the outbreak revealed that some burgers had cool and bloody areas; one spot in one burger had reached only 110 degrees.[38]) Prior to 1993 it is difficult to find evidence that the USDA did anything to attack the problem. Its own early tests had found the microbe to be only rarely present on beef, but according to a USDA source, the agency was not using the sorbitol-MacConkey culture medium, without which isolating the microbe is difficult.

Proposed regulations have met fierce industry resistance. Various sectors of the food industry fought attempts to raise the cooking temperature on precooked meat patties, despite an outbreak in a school cafeteria. The industry opposed warning labels on meat, including safe handling and cooking instructions, and even sought and obtained an injunction from Federal Judge James R. Nowlin, in Austin, Texas, that held up labels by five months. Two children died in Texas from HUS in the weeks following that decision.

Although the speed of meat-processing operations is implicated in the contamination of meat with fecal material, lines in the slaughterhouses accelerated even more during the deregulation frenzy in the 1970s and 1980s. Under experimental programs the number of inspectors was reduced. Unsanitary conditions were allowed to continue, and repeated violations failed to lead to USDA action. Some inspectors complained of being unable to do their jobs. A pilot program called the Streamlined Inspection System (SIS) brought forth horrifying descriptions of foul and unsanitary conditions from whistle-blowing meat inspectors. Line speeds often reached four hundred cattle per hour; inspectors were asked to check as many as thirty livers per minute and found the task impossible; grossly infected cattle with such conditions as urine-filled bellies, peritonitis, pneumonia, and measles (tapeworm) were slipping past inspectors into the food chain. Inspectors complained that they could no longer see the inside of the carcass and had lost the authority to trim to expose signs of disease. Cow heads, which are trimmed to provide meat for hamburger, were getting through with regurgitated food oozing out, and "fecal contamination and other filth are getting out of control." Plants that had once had clean floors were, under the new system, "full of guts, urine, feces and general muck, sometimes to the point of being so slippery it's dangerous to walk." "Much of what USDA calls wholesome today," said one affidavit, "would have been condemned in 1984, and it wouldn't have been a close call. Meat whose disease symptoms previously would have forced it to be condemned, or at most approved for dog food, now gets the USDA seal of approval for consumers."[39]

The SIS pilot program was eventually dropped, but the 3.5-percent rate of *E. coli* contamination in hamburger that industry itself found should have come as no surprise.[40] Many critics charged that USDA decision-making was doing more to please industry than to ensure food safety.

## The USDA Response

When informed by the Washington State Department of Health of the connection between the Jack-in-the-Box outbreak and USDA-inspected meat, Jill Hollingsworth, assistant administrator of the Food Safety and Inspection system (FSIS), told the health department's Charles Bartleson, "We will take no action because this meat does not violate USDA standards." To which Bartleson replied, "I thought you guys were in the public health business."[41] He could be forgiven for his confusion.

Dr. Russell Cross was then the FSIS chief and much admired by industry. In a briefing memo to the new secretary, Cross apprised Espy of the department's position on *E. coli* O157:H7, echoing Hollingsworth's line. "Under current regulations," he said, "the presence of bacteria in raw meat, including *E. coli* O157:H7, although undesirable, is unavoidable, and not cause for condemnation of the product. Because warm-blooded animals naturally carry bacteria in their intestines, it is not uncommon to find bacteria on raw meat."[42]

Cross failed to mention to Espy that muscle meat is, in fact, sterile, and that fecal material transferred the bacteria to raw meat during the USDA-inspected slaughtering process. He would repeat this agency disclaimer that the meat was not contaminated several times over the next few weeks as public Senate hearings addressed the outbreak. It was frustrating, he said, that so little was known about the pathogen.

In fact, medical researchers and epidemiologists knew a great deal. The information was as near as the nearest medical library in the great number of journal articles that had been published since 1983. But there seemed to have been a breakdown in communication between the USDA and the CDC— between the world of human health and the world of animal health. That sharp separation was no longer appropriate. The two were drawing ever closer.

The USDA position on microbial contamination was not new. It was based on a decision by the District of Columbia Court of Appeals in the case of *American Public Health Association* v. *Butz* (1974), which said that Congress had not intended that inspections would include "microscopic examinations." In 1987, when the pathogen was discovered in commodity meat, causing four deaths in two institutions in Utah, the USDA decided that the meat need not be sampled or recalled, because it was "safe" if "cooked properly," and that sampling or recalling it "would create more problem [*sic*] for the agency, especially if the press learns of the FSIS action on the product that has already been released by the state of Utah."[43] The meat from the contaminated lot was distributed by the state to other facilities.

The Jack-in-the-Box outbreak brought wide media attention. If Espy did not learn of the particular virulence and the source of *E. coli* O157:H7 from his undersecretary in charge of food safety, he heard of it elsewhere. He was concerned about the agency's reputation. Consumers would be unlikely to appreciate the subtlety of the argument that the meat wasn't, by definition, contaminated, when children were dying. In the following weeks and months the secretary would hire two hundred more inspectors, institute a policy of zero tolerance for fecal material on carcasses, and order surprise inspections of slaughterhouses and meatpacking facilities—although most reasonable members of the public could be forgiven for wondering why these steps had not been taken before. Only resistance from within the department kept Espy from doing more, aides said.

On March 2, thirty-five meat inspectors wrote Espy to say that they expected retaliation from plant managers for attempting to carry out zero tolerance. They explained that Cross had "almost immediately gutted" the standard by telling a plant official that the ruling applied only to "obvious" fecal contamination and inviting him to report any inspectors applying "knee-jerk" interpretations of zero tolerance. "We are now in a position where FSIS leaders may have invited industry to prepare a hit list of inspectors whose crime is following the agency's own orders," the inspectors wrote.[44]

"We are not talking about anything punitive," Cross would explain to the *Federal Times*. "I just wanted to know if there were any inconsistencies."[45]

A spokesperson for the Government Accountability Project, which represents whistle-blowers, said that Secretary Espy was being undermined "by a holdover FSIS bureaucracy. While Secretary Espy says the right thing in public, Dr. Cross undercuts him by doing the wrong thing in private."[46] Eventually Espy would decide that he was being ill served by Dr. Cross, and Cross and two others would resign.

Charges that Espy had acted unethically by accepting gifts of travel and entertainment from Tyson Foods, the giant Arkansas chicken producer, had begun to surface and had the secretary's staff on edge. The Tyson-Clinton connection, even as the Whitewater tale developed, made the story irresistible to journalists. Then from within the department came charges from Wilson S. Horne, former head of inspection operations, that Espy's chief of staff had suddenly stopped work on regulations to reduce pathogens in chicken. The charge was denied.

Espy loyalists, who had from the beginning found the agency's response to food safety weak, thought the attacks were originating with the beef industry. In fact, beef representatives have long felt that poultry has the regulatory advantage. Under USDA rules, poultry processors could add water weight, and did (8 percent); chlorine could be added to the water; irradiation had been approved for chicken; fecal contamination was allowed, as were pathogens; and in ground chicken products, lungs, sex glands, and skin could be a part of the mix. The beef industry wanted a level playing field.[47]

Battle lines were drawn. In a *Los Angeles Times* interview in July, Espy publicly said that he had the impression from officials in his department that they knew about the deaths in the 1993 outbreak but weren't concerned. "They considered the deaths acceptable," he told the *Times*.[48] Jill Hollingsworth, who'd been so quick to deny agency responsibility to Bartleson, told employees in a conference call that she couldn't imagine how the secretary "ever walked away with [that] impression."[49] An anonymous letter purporting to be from some USDA employees to the president and asking for his help came in August. In it Espy was accused of being "desperate, malicious, and slanderous," and of using "food safety as his personal ticket to fame." The letter continued: "It is not acceptable for people to die from the food they eat . . . but it happens."[50] The writers said they feared retaliation if they revealed themselves. There was no way to tell whether the document was authentic.

Those in the department who knew what Espy was complaining about did their own leaking. Food safety advocates had long suggested that rapid microbial tests were needed to detect the presence of microbes and had pressed the USDA to develop one. In August, Dan Laster, director of the U.S. Meat Animal Research Center in Clay Center, Nebraska, wrote a memo detailing what he had previously revealed to a reporter, that several senior-level FSIS staff had stymied his efforts to develop the test. They had, he said, "tried very hard to prevent us from moving forward in any kind of timely manner to evaluate the Rapid Test under in-plant conditions."[51] The "stonewalling" was thought to have been an effort to embarrass Secretary Espy.

In fact, the attacks on Espy in the press were increasing. He knew time was short. In August he replaced Cross at the FSIS with Michael R. Taylor, who'd won praise from consumer advocates while at the Food and Drug Administration. Taylor stood at Espy's side as legislation proposing the new Pathogen Reduction Act was introduced to Congress. It would have given the USDA authority to recall products (now it can only withhold its stamp of inspection and encourage industry to recall) and impose fines, and it would have established microbial limits beyond which a product could be considered contaminated. It was a bold move by a wounded secretary. Heat was growing in the charges of misconduct. Days later he would announce his resignation and the Pathogen Reduction Act would, with advice from industry, consumers, and Congress, begin to evolve.

Taylor, too, planned to move quickly and decisively to do something about meat safety. His most dramatic gesture came on September 29, 1993, in an address to the American Meat Institute. When he got to the subject of *E. coli* O157:H7, the members of his audience, who'd been listening with half an ear, began frantically scribbling notes. The bacterium, he said, would from now on be considered an adulterant on meat. The shock waves could be felt through the audience. This was a historic change. Taylor went on. Because *E. coli* O157:H7 was unusually virulent—some researchers had suggested that as few as ten microbes could cause illness—any amount of it would render the product unsafe, and the agency did have the authority to hold product that was clearly unsafe. Not only that, the agency would immediately begin testing for *E. coli* O157 at the retail and "grinder" level—although it would take only five thousand samples.

It was perhaps the single most important change in USDA history—the agency would do something it had never done, test for a microbe and hold the product if it was found—and yet the speech was not routinely released, as most official addresses are. Copies had to be requested. And the Associated Press delayed reporting the story for several days, for reasons that are not clear.

Taylor's speech marked an about-face that left the industry outraged. If *E. coli* O157:H7 could be considered an adulterant, so might *Salmonella* or *Listeria* or *Campylobacter* or any of the other pathogens that were getting into meat during slaughter and production. The industry protested vehemently in the days following, threatening to file a lawsuit, but Taylor was unmoved. He tossed the industry a bone: It could bypass the usual process for changing regulations and use organic acid sprays on the carcasses, although processors would still be required to trim. One study had shown acid sprays to be effective, and another that they were ineffective; spraying was cheaper, however, than the real solution of trimming away fecally contaminated meat. And besides, spraying added water weight—if "only one-half of one percent," in the words of an industry representative. But in a large-scale operation, a minute percentage can mean huge increased profits.

Taylor's gesture did not mollify the industry. On November 1, the American Meat Institute (AMI), the Food Marketing Institute (FMI), the Associated Grocers of America, and four other groups filed a lawsuit seeking a permanent injunction against the USDA to halt its testing of hamburger for

O157. Their interesting rationale: Testing might lead consumers to think their meat was safe and to ignore the cooking and handling warnings that the industry had fought so hard to keep off meat. The industry took its suit to the same Judge Nowlin who had ruled in its favor before. The AMI-FMI lawsuit said the USDA failed to follow proper rule-making procedures in making the change. It used the agency's own inaction since the microbe first caused trouble in 1982 as evidence the *E. coli* O157:H7 neither represented adulteration nor presented an emergency situation. The industry saw in the new definition a frightening potential for lawsuits. (One industry lawyer advised producers not to test their own meat, or to stop if they were already doing so, to avoid liability.)

The USDA was not surprised. It had anticipated something of the sort. But Taylor's staff had gambled that the weight of public opinion now favored food safety. And indeed, meat industry representatives were surprised when, meeting in Chicago to discuss foodborne illness three days after filing the suit, they were met by protesters from STOP bearing signs such as "Industry obstructs again" and "Testing = Accountability." The protests made some major newspapers.

The industry has argued for the past few years that USDA inspections are outdated. Inspectors, they argue, should abandon the traditional "organolyptic" methods, which use touch, sight, and smell, for "science-based inspection." Why, then, would the industry oppose the "science-based" microbial testing the USDA has begun?

The industry argues that testing at the retail level would not prevent the microbe from reaching the public—a negative test, as one industry scientist put it, "only tells you that you don't have it in that sample"—but might damage the reputation of the product. Grocers, for instance, resent risking their reputations when contamination probably occurred at the slaughterhouse level. There is some merit to that argument. But testing at an earlier point in the process can give an idea as to what and how many pathogens are present. It was that tack the USDA would pursue.

The "science-based" inspection the industry prefers is called the hazard analysis and critical control points system (HACCP, pronounced "hasip"), which establishes where dangers lurk in the process and monitors those points. HACCP could help; it has shown its effectiveness in other industries. In a sardine plant, for instance, the fish can be visually checked as they go into the can and are sealed, but the critical control point is the "retort," where the sealed can is subjected to intense steam heat for a certain length of time.

The HACCP system requires that the time and temperatures be recorded and checked. But there is one essential difference between sardines and meat: Unlike ground beef, which generally carries the label of the store on the package rather than the producer or processor, accountability is built into the sardine can. And there is an easily identifiable critical control point. No pathogenic microbes (that we presently know about) can survive that process. But HACCP remains unproven with raw meat. The critical control points at which screening or testing might control this tenacious microbe are only now being identified, and there is no process that can render raw meat totally free of microbes. (Even irradiation leaves a few hearty microbes such as *Clostridium botulinum* and *Clostridium perfringens*.)

The USDA feared that Judge Nowlin, who had confessed to being a cattleman and wondered aloud to the lawyers whether that might constitute a conflict of interest, would rule from the bench on the AMI-FMI lawsuit. He did not. When he finally ruled on December 13, he declined to stop testing. The USDA had won a round. A furious meat industry promised to fight on, but in fact it would back off publicly. There wasn't a lot of good press to be gained by fighting food safety in the open. The industry would confine its considerable influence to getting the best deal it could from the upcoming HACCP regulations.

HACCP, in the version proposed by the AMI, would do more than systematize slaughter. It would have industry inspectors replace government inspectors, who would remain on the job but would shift from looking at product and process to looking at paper. It's the deregulatory aspects of HACCP that make it so appealing to the industry. The USDA began preparing new regulations that would include HACCP but also microbial testing. "We have a vision of HACCP that's going to be different from industry's vision," laughed Taylor.[52]

(The legislation would take time. Behind the scenes battle plans were drawn as industry, medical researchers, public health officials, and the consumer food safety organizations that now had a seat at the table vied to see who would have more influence. It would be 1996 before the legislation was finally introduced, and it would be a compromise version.)

*E. coli* O157 is an emergency situation, the USDA had argued in its brief response to the AMI-FMI lawsuit. The agency hoped that testing for the microbe would prompt the industry to take action. The aim, says Taylor, is to increase accountability. But in 1996, after two years of testing, the USDA had come up with only five positives—even as outbreaks and sporadic cases continued to occur. By contrast, David Acheson of Tufts University had devised a test to look not for the microbe but for the Shiga toxin produced by both O157 and some non-O157 *E. coli*. When he tested it on sample packages of raw hamburger, his baseline positive rate was 25 percent. He went back to these positives and confirmed them by finding the microbe. Any reasonable observer might wonder whether something wasn't wrong with the USDA's testing procedure.

Testing was not an entirely new thought. Florida had begun testing for the microbe fourteen months earlier than the USDA. A few days after the lawsuit was filed, Florida found it. Montfort, Inc., of Greeley, Colorado, could recall nine thousand pounds of *E. coli* O157:H7–contaminated hamburger meat.

While thorough cooking is assumed to destroy the bacteria—a fact that has led the meat industry to call the consumer or food preparer the ultimate "critical control point"—the USDA now argues that many Americans are accustomed to eating their burgers rare or pink in the middle and are thus vulnerable. Not only that, cross-contamination in the kitchen can present a huge problem. In one case a teenager was preparing his own supper and began cooking a burger. He cooked it thoroughly. His brother came in while the burger was still cooking and tossed his own into the skillet, turning it and mashing it down with a spatula. The first boy used the same spatula to remove his burger. It was that boy who became ill.

The microbe has continued to demonstrate its adaptability. An outbreak of *E. coli* O157–associated illness in Washington State and California affecting eighteen people was associated with cured salami previously thought to be safe. When the manufacturer recalled ten thousand pounds, another segment of the food industry was jolted into the new O157 reality. The pathogen is more acid-resistant than most *E. coli*, and the traditional fermentation process hadn't stopped it. Since the "just cook it" message to consumers doesn't fit a product meant to be consumed as is, the cured-sausage industry is deeply troubled. But it should have anticipated the problem. Studies published in 1992 by Michael Doyle of the University of Georgia, a scientist who has researched *E. coli* O157:H7 for years, had indicated that the microbe survives both fermenting and drying. "Some folks in the industry were well aware of it," he says.[53]

But one food-safety advocate has advice for consumers, who may by now feel battered and vulnerable. Mary Heersink of STOP suggests putting a lump of any newly purchased ground beef and the receipt into a plastic bag and freezing it, thus creating what she calls a "frozen insurance policy and instant accountability" for that moment when the hospital asks whether you have any left. The thought may well have sent a chill through the industry.

Since the Jack-in-the-Box (western states) outbreak, there have been, according to STOP, which keeps an unofficial tally with their 800 number, more than two hundred outbreaks. Since the CDC estimates that there are 250–500 deaths a year from O157, and since ground beef is the primary vehicle, it is a real possibility that every day in this country someone dies from eating a hamburger, a thought that lends a new dimension to one of America's cultural icons.

The CDC had tried to trace the meat in the western states outbreak back to its source, but large-scale slaughtering and mass-distribution made doing so virtually impossible. The records simply stopped, and "no one could determine exactly what meat went into any lot," said Patricia Griffin. The agency did discover two points in the slaughter and deboning process that didn't look good. Some of the slaughterhouses had used "bed slaughter": laying the animal on a bed of rails inches from the floor, which "is often covered with feces and dirt; the animal is then skinned by hand." The other problem seemed to be the boning plant. "Meat is removed from the bones on a long table. A conveyer belt carries the meat to one end. The belt is sanitized once daily."[54] The boning belt was ideal for cross-contamination.

Griffin was able to establish a probable chain of events that resulted in the outbreak.

> A farm supplied cattle carrying *E. coli* O157:H7. A boning plant then supplied the contaminated meat to one hamburger-making plant. The hamburger-making plant mixed contaminated meat with other meat. Contaminated hamburgers were made in huge quantity on 2 successive days several weeks before the outbreak and were frozen for wide distribution. The final defense was lost when the hamburgers were insufficiently cooked in more than 90 Chain A restaurants in 4 states. As a result, a single strain of *E. coli* O157:H7 infected over 700 people.
>
> In conclusion, *E. coli* O157 colonization of cattle results in human illness and death. Current slaughtering methods can result in fecal contamination of meat. Central processing methods that mix contaminated meat with other meat can result in many contaminated hamburgers. Fast-food chains can cause widespread out-

breaks because they can have many restaurants using a uniform cooking method on meat from a common source.[55]

Griffin was optimistic. "The impact of this outbreak has been impressive. *E. coli* O157:H7 became a household word. Food safety became a hot topic. . . . The impact on the U.S. Department of Agriculture has been unprecedented. . . . Congress also awakened to food safety," she wrote. But it was 1994, and while Congress had been forcibly awakened to food safety, its actions after the November elections would be disappointing. And while it may have seemed to Griffin, from her perspective, that surely everyone knew about *E. coli* O157:H7 and the dangers of rare burgers or cross-contamination in the kitchen, she couldn't have been more wrong. The pathogen might have received massive publicity, but much of it had been inaccurate and misleading; it had minimal impact on a public conditioned to hyperbole.

In the summer of 1996 an expensive restaurant in Waitsfield, Vermont, would outline in its menu how it prepared its burgers. Rare was defined as pink and warm inside, medium was pink but hot. Well done, brown all the way through, came with a warning: "We are not responsible." It should have been the other way around. The message would be slow to penetrate the culture and then, at some levels, would be generally scoffed at. One more health scare. People were tired of hearing it. And most thought they were magically immune.

Vital questions remain unanswered: Where did the organism come from, and why was it in cattle? No one was sure, but clues as to why it was getting on meat pointed to a food safety strategy gone wrong. Slaughterhouses had assumed they could lower the level of contamination on meat if they emptied the intestines of animals by withdrawing food prior to killing them. But a number of studies would make this appear counterproductive.[56] And there were other changes that USDA researchers thought might be contributing.

Gregory Armstrong, Jill Hollingsworth, and Glenn Morris, Jr., of the FSIS identified these factors in an article that appeared in 1996 in *Epidemiologic Reviews*.[57] Over the past twenty years the beef cattle industry, they wrote, has become ever more concentrated at the feed lot level, decreasing from 121,000 feedlots in 1970 to 43,000 in 1988. The largest of these fatten as many as 16,000 animals at one time. The same trend can be seen in the dairy industry, which supplies spent animals for hamburger meat. While there were 600,000 dairy farms in 1955, by 1989 there were only 160,000, and the number of farms with milk cows decreased as well, from 2,800,000 to 205,000.

Changes in how animals are raised—from what they eat to how they eat—may also have contributed to the growing presence of this new pathogen in cattle. Computerized feeding has been linked to animals found to harbor *E. coli* O157:H7 and the feeding of antibiotics, a practice that began in the mid-1970s, parallels the emergence of this pathogen. Perhaps the spreading of manure slurry on fields may play a role, they suggested, as may, ironically, the success in reducing another disease in animals. Immunity to brucellosis, brought on by an animal's response to infection, can provide immunity to *E. coli* O157:H7, and there seems to be more O157 infection in areas, such as the Northwest, that have been successful in reducing the level of brucellosis in cows.

But changes in the way meat is processed are having a profound effect. The USDA researchers estimated that with the mass production now standard, one infected animal could contaminate 16 tons of hamburger meat. And the size of hamburger lots is, in turn, related to concentration and vertical integration in the meat industry. The four largest meat-packing firms increased their market share from 22 percent in 1977, to 32 percent in 1982, to 54 percent in 1987. Even more concentrated is the boxed beef industry. In 1987, 80 percent of the market share went to the four largest firms.

At the same time, consumers are changing their eating practices. While consumption of hamburgers has actually decreased since 1976, eating at fast-food restaurants increased 224 percent from 1967 to 1982. While these burgers are not intentionally undercooked (although it clearly happens), when they have a choice, either at home or in non-fast-food restaurants, 23–25 percent of Americans prefer their beef rare.

While no one change played the key role, it is likely that each change contributed. "In particular," say Armstrong, Hollingsworth, and Morris, "big may not always be better. Consolidation of the industry, widespread movement of cattle, increased use of large production lots for products such as hamburger, may all have played a role in the process—and may provide a setting in which other 'new' pathogens can rapidly move into human populations."[58]

At the molecular level the evidence seemed to show that O157:H7 had emerged recently when a strain of *E. coli* related to the enteropathogenic series picked up the ability to produce the toxin from *Shigella* by horizontal transfer—bacteria can share genetic information in DNA in a direct transfer process called conjugation. While chromosome transfer is relatively rare, plasmid transfer occurs frequently and rapidly among some bacteria.

Said Marguerite Neill,

> Judicious reflection on the meaning of this finding suggests a larger significance—that *E. coli* O157:H7 is a messenger, bringing an unwelcome message that in mankind's battle to conquer infectious diseases, the opposing army is being replenished with fresh replacements. . . . We have no cogent explanation for why this pathogen has appeared and we do not know whether we are fostering its dissemination. We have no detailed understanding of pathogenesis at the cellular level, leaving us without a scientific basis to design treatment strategies or prevent complications. We do not know what control measures work or where to apply them. We do not yet have a conceptual approach to *E. coli* O157:H7 which incorporates a comprehensive public health outlook with practical cost-effective control measures. It is sobering to note that all the features of the 1993 Washington State outbreak were already known by 1984.

That is, *E. coli* O157:H7 is a high-grade pathogen that needs only a few microbes to cause disease and is transmitted by a high-volume food item, she said, "whose preparation contains a compositional and thermal Achilles heel," and it is served to a "target audience [children] most at risk for complications of illness."[59]

## Notes

1. Mary Heersink, in-person interview, Dothan, Alabama, and Washington, D.C.; 1993, 1994, 1995, 1996.
2. Jeffrey Tennyson, *Hamburger Heaven: The Illustrated History of the Hamburger* (New York: Hyperion, 1993), pp. 20–24. (Note: Much of what I know about the history of the hamburger can be credited to Tennyson.)
3. Ibid., pp. 58–85.
4. Ibid., p. 80.
5. Richard Hebert, telephone interview, 1993.
6. Lee Riley, telephone interview, 1993. (Note: See also L. W. Riley, R. S. Remis, S. D. Helgerson et al., "Hemorrhagic colitis associated with a rare *Escherichia coli* serotype," *New England Journal of Medicine* 308 (1983): 681–85.)
7. Joy Wells, in-person interview, Atlanta, 1993.
8. Peggyann Hutchinson, telephone interview, 1993.
9. Lynn Turner, telephone interview, 1993.
10. Robert Remis, telephone interview, 1993.

11. Riley et al., p. 682.
12. Steve Steinberg, "Undercooked Burger Linked to Disease," *Miami Herald*, October 7, 1982.
13. Ibid.
14. Nathaniel Sheppard, Jr., "McDonald's Linked to 47 Stomach Disorder Cases," *New York Times*, October 8, 1982.
15. Edward Edelson, "A Big Mac Attack?" *New York Daily News*, October 8, 1982.
16. Edward Edelson, "McDonald's Gets Break on Ailment," *New York Daily News*, October 9, 1982.
17. Allen Hallmark, "Illness Victims Ate at McDonald's," *Medford Mail Tribune*, October 9, 1982.
18. V. C. Gasser, E. Gautier, A. Steck, R. E. Siebnmann, R. Oechslin, "Hämolytischurämmische Syndrome: Bilaterale Nierenrindennekrosen bei akuten erworbenen hämolytischen Anamien," *Schweiz Med Wochenschr* 65 (1955): 905–9.
19. Brian T. Steele, telephone interview, 1993.
20. Mohamed A. Karmali, telephone interview, 1993.
21. C. A. Ryan, R. V. Tauxe, G. W. Hosek et al., "*Escherichia coli* O157:H7 Diarrhea in a Nursing Home: Clinical, Epidemiological, and Pathological Findings," *Journal of Infectious Diseases* 154 (1986): 631–38.
22. Caroline A. Ryan, telephone interview, 1994.
23. Robert V. Tauxe, in-person interview, Atlanta, 1993.
24. A. O. Carter, A. A. Borczyk, J. A. K. Carlson et al., "A Severe Outbreak of *Escherichia coli* O157:H7–associated Hemorrhagic Colitis in a Nursing Home," *New England Journal of Medicine* 317 (1987): 1496–1500.
25. S. M. Ostroft, P. M. Griffin, R. V. Tauxe et al., "A Statewide Outbreak of Escherichia coli O157:H7 Infections in Washington State," *American Journal of Epidemiology* 132 (1990): 239–47.
26. Patricia Griffin, telephone interview, 1997.
27. In a letter from Kenneth C. Clayton, Acting Assistant Secretary for Marketing and Inspection Services, Department of Agriculture, to MS. Kathi Smith Allen on April 21, 1993, Clayton wrote, "CDC first reported cases of hemolytic uremic syndrome, a urinary tract infection caused by the bacteria, in 1982."
28. Anonymous, "Need for USDA Beef Patty Rule Questioned: Additional Research Asked," *Food Chemical News*, January 20, 1989.
29. Robert Galler, prepared and delivered statement, Symposium on Meat Inspection, Washington, D.C., September 1993.
30. Arthur O'Connell, prepared and delivered statement, Symposium on Meat Inspection, Washington, D.C., September 1993.
31. Wendy Feik Pinkerton, C. J. Reynold, "Washington State *E. coli* O157:H7 Outbreak," Memorandum to State Beef Council Executives, National Live Stock and Meat Board, January 19, 1993.
32. Steve Bjerklie, telephone interviews, 1996.
33. Edward Penble, "USDA Slow to Respond to Food Poisoning," *Seattle Post-Intelligencer*, January 21, 1993.

34. Ibid.

35. Dorothy Dolan, prepared and delivered statement, Symposium on Meat Inspection, Washington, D.C., September 21, 1993.

36. Diane Nole, prepared and delivered statement, Symposium on Meat Inspection, Washington, D.C., September 21, 1993.

37. Robert V. Tauxe, in-person interview, Atlanta, 1993.

38. D. Maxson, employee, Foodmaker (parent company of Jack-in-the-Box), "Patty Temperature Tracking Form: Burger Patty Observations," Store 7211, February 10, 1993.

39. Government Accountability Project, "Summary of 1990 Whistle-blowing Disclosures on USDA's Proposed Streamlined Inspection System—Cattle," Governmental Accountability Project, Washington, D.C.

40. James L. Marsden, Issues Briefing "Issue: *E. coli* O157:H7," American Meat Institute, July 1992. Marsden wrote: "Dairy cattle are a major reservoir of *E. coli* O157:H7. The organism resides in their intestinal tracts, and they shed it in feces. Slaughter and milking procedures can contaminate meat and milk with *E. coli* bacteria, but because milk is generally pasteurized and meat typically is cooked prior to ingestion, the presence of *E. coli* bacteria in general is not considered a public health hazard . . . Foodborne illness cases associated with *E. coli* O157:H7 have been linked to undercooked ground beef and, to a lesser extent, improperly pasteurized or raw milk. Supermarket surveys of fresh meats and poultry revealed *E. coli* O157:H7 in 3.5 percent of ground beef, 1.5 percent of pork, 1.5 percent of poultry and 2.0 percent of lamb."

41. Charles Bartleson, personal notes, January 26, 1993.

42. H. Russell Cross, Informational Memorandum for the Chief of Staff, *E. coli* O157:H7 Outbreak," January 29, 1993.

43. Anonymous Report on "FSIS Meeting on EP case no. 7078, Edid. case no. 870290," USDA Document, August 28, 1987.

44. Meg Walker, "Meat Inspectors Fear Retaliation," *Federal Times*, March 29, 1993.

45. Ibid.

46. Ibid.

47. USDA, "Poultry Products Produced by Mechanical Deboning and Products in Which Poultry Products Are Used," *Federal Register*, March 33, 1994.

48. *Los Angeles Times*, July 29, 1994.

49. Sharon LaFraniere and Gay Gugliotta, "The Promise and Puzzle of Mike Espy," *Washington Post*, September 28, 1994.

50. A copy of the anonymous letter to the president was obtained from a reliable source; it was also referenced in a *Washington Post* article, which bolsters its authenticity.

51. D. B. Laster, Central Director, Clay Center, Nebraska, "Information on Rapid Bacterial Test for Meat and Poultry," Memorandum to R. D. Plowman, Acting Assistant Secretary, USDA, and P. Jensen, Acting Assistant, USDA, August 11, 1994.

52. Michael R. Taylor, telephone interview, 1994.

53. Michael Doyle, telephone interview, 1994.

54. P. M. Griffin, B. P. Bell, P. R. Cieslak et al., "Large Outbreak of *Escherichia coli* O157:H7 Infections in the Western United States: The Big Picture," *Elsevier Science B.V.*, 1994, pp. 7–12.

55. Ibid., p. 11.

56. M. A. Rasmussen, W. C. Cray, T. A. Casey, and S. C. Whipp, "Rumen Contents as a Reservoir of Enterohemorrhagic *Escherichia coli*," *FEMS Microbiology Letters* 114 (1993): 79–84.

57. G. L. Armstrong, J. Hollingsworth, and G. Morris, Jr., "Emerging Foodborne Pathogens: *Escherichia coli* O157:H7 as a Model of Entry of a New Pathogen in the Food Supply of the Developed World," *Epidemiologic Review* 19 (1996): 1–23.

58. Ibid., pp. 15–16.

59. M. Neill, "*E. coli* O157:H7 Time Capsule: What Do We Know and When Did We Know It?" Lecture presented at the symposium on Foodborne Microbial Pathogens. Annual Meeting of the International Association of Milk, Food and Environmental Sanitarians, Atlanta, Georgia, August 2–4, 1993.

# The United States Food Safety System      20

U.S. FOOD AND DRUG ADMINISTRATION

*The various agencies of the United States that are responsible for the safety of Americans' food typically claim that the United States has the safest food system in the world. Typical among such statements is the one included here from the U.S. Food and Drug Administration (FDA).*

*This document gives a reasonable overview of how the American government tries to ensure the safety of food. What it doesn't say is how closely the various agencies are connected to huge agribusinesses with links to key politicians. The document indirectly reveals the fragmented, piecemeal nature of this system, with four different agencies having separate, sometimes overlapping, authority over some areas (which leads to conflicts about responsibility). Reading this document, one realizes why some critics of the current system have called for a new "Czar of Food," a cabinet-level position that would put all aspects of food production (growing, subsidies, safety, and development) under a single, consistent agency.*

*The document is also somewhat self-serving and makes things appear better and safer than they are. In discussing transparency or openness to public scrutiny, the document makes it appear that the United States has an adequate system for alerting the public of food risks. Yet the meat industry is not legally required to inform the public of many risks and we do not know how many times it has chosen not to do so.*

*Hundreds of thousands, maybe millions, of people suffer incidents of food intolerance (not to be confused with the more serious food allergy) every year. Almost all of these go unreported because only cases requiring a physician's attention are reported to the Centers for Disease Control. Hence, just how safe the American food system is, especially in its modern, industrial guise, remains unclear.*

---

From U.S. Food and Drug Administration, "United States Food Safety System," downloaded from the FDA website (www.foodsafety.gov) on May 16, 2001.

# I. Synthesis: The United States Food Safety System

THE UNITED STATES CONSTITUTION prescribes the responsibilities of the government's three branches: executive, legislative and judicial, which all have roles that underpin the nation's food safety system. Congress, the legislative branch, enacts statutes designed to ensure the safety of the food supply. Congress also authorizes executive branch agencies to implement statutes, and they may do so by developing and enforcing regulations. When enforcement actions, regulations, or policies lead to disputes, the judicial branch is charged to render impartial decisions. General U.S. laws and statutes and Presidential Executive Orders establish procedures to ensure that regulations are developed in a transparent and interactive manner with the public. Characteristics of the U.S. food safety system include the separation of powers among these three branches and transparent, science-based decision-making, and public participation.

The U.S. food safety system is based on strong, flexible, and science-based federal and state laws and industry's legal responsibility to produce safe foods. Federal, state, and local authorities have complementary and interdependent food safety roles in regulating food and food processing facilities. The system is guided by the following principles: (1) only safe and wholesome foods may be marketed; (2) regulatory decision-making in food safety is science-based; (3) the government has enforcement responsibility; (4) manufacturers, distributors, importers and others are expected to comply and are liable if they do not; and (5) the regulatory process is transparent and accessible to the public. As a result, the U.S. system has high levels of public confidence.

Precaution and science-based risk analyses are long-standing and important traditions of U.S. food safety policy and decision-making. U.S. food safety statutes, regulations, and policies are risk-based and have precautionary approaches embedded in them.

The agencies' well-qualified science and public health experts work cooperatively to ensure the safety of U.S. food. Scientists from outside government are regularly consulted to provide additional recommendations regarding technical and scientific methods, processes, and analyses used by regulators. The cutting-edge science that informs U.S. regulators is routinely shared internationally through interactions with organizations like the Codex Alimentarius Commission, World Health Organization, the Food and Agriculture Organization, and the International Office for Epizootics.

The U.S. routinely and effectively deals with technological advances, emerging problems, and food safety incidents. It is enhancing early warning systems about pathogens in food. The legislation granting authorities to agencies generally enables them to revise regulations and guidance consistent with advances in technology, knowledge, and need to protect consumers.

U.S. food agencies are accountable to the President, to the Congress which has oversight authority, to the courts which review regulations and enforcement actions, and to the public, which regularly exercises its right to participate in the development of statutes and regulations by communicating with legislators, commenting on proposed regulations, and speaking out publicly on food safety issues.

## II. United States Food Safety System

### Introduction

The U.S. food safety system is based on strong, flexible, science-based laws and industry's legal responsibility to produce safe foods. Coordinated interactions among federal authorities having complementary and interdependent food safety missions, in partnership with their state and local government counterparts, provide a comprehensive and effective system. The implementation of the statutes and the food safety system over many years has resulted in very high levels of public confidence in the safety of food in the U.S.

Principal federal regulatory organizations responsible for providing consumer protection are the Department of Health and Human Services' (DHHS) Food and Drug Administration (FDA), the U.S. Department of Agriculture's (USDA) Food Safety and Inspection Service (FSIS) and Animal and Plant Health Inspection Service (APHIS), and the Environmental Protection Agency (EPA). The Department of Treasury's Customs Service assists the regulatory authorities by checking and occasionally detaining imports based on guidance provided. Many agencies and offices have food safety missions within their research, education, prevention, surveillance, standard-setting, and/or outbreak response activities, including DHHS's Centers for Disease Control and Prevention (CDC) and National Institutes of Health (NIH); USDA's Agricultural Research Service (ARS); Cooperative State Research, Education, and Extension Service (CSREES); Agricultural Marketing Service (AMS); Economic Research Service (ERS); Grain Inspection, Packers and Stockyard Administration (GIPSA); and the U.S. Codex office; and the Department of Commerce's National Marine Fisheries Service (NMFS).

The FDA is charged with protecting consumers against impure, unsafe, and fraudulently labeled food other than in areas regulated by FSIS. FSIS has the responsibility for ensuring that meat, poultry, and egg products are safe, wholesome, and accurately labeled. EPA's mission includes protecting public health and the environment from risks posed by pesticides and promoting safer means of pest management. No food or feed item may be marketed legally in the U.S. if it contains a food additive or drug residue not permitted by FDA or a pesticide residue without an EPA tolerance or if the residue is in excess of an established tolerance. APHIS' primary role in the U.S. food safety network of agencies is to protect against plant and animal pests and diseases. FDA, APHIS, FSIS, and EPA also use existing food safety and environment laws to regulate plants, animals, and foods that are the results of biotechnology.

## A. Laws and Implementing Regulations

The three branches of U.S. government—legislative, executive, and judicial—all have roles to ensure the safety of the U.S. food supply. Congress enacts statutes designed to ensure the safety of the food supply and that establish the nation's level of protection. The executive branch departments and agencies are responsible for implementation, and may do so by promulgating regulations, which the U.S. publishes in the *Federal Register* and which are also electronically available. Characteristics of the U.S. food safety system are the separation of powers and science-based decision-making. Agency decisions under U.S. food safety laws can be appealed to the courts which are empowered to settle such disputes.

Food safety statutes enacted by Congress provide regulatory agencies with broad authority but also set limits on regulatory actions. The statutes are drafted to achieve specific objectives. Food safety agencies then develop regulations that give specific direction and establish specific measures. When new technologies, products, or health risks must be addressed, agencies have the flexibility to revise or amend regulations generally without need for new legislation. Agencies are able to maintain their state-of-the-art scientific methods and analyses because changes of this type can be made at the administrative/technical level.

Major U.S. food safety authorizing statutes include the Federal Food, Drug, and Cosmetic Act (FFDCA), the Federal Meat Inspection Act (FMIA), the Poultry Products Inspection Act (PPIA), the Egg Products Inspection Act (EPIA), Food Quality Protection Act (FQPA), and Public Health Service Act.

Procedural statutes, which regulatory agencies must follow, include the Administrative Procedure Act (APA), the Federal Advisory Committee Act (FACA), and the Freedom Of Information Act (FOIA). The APA specifies requirements for rulemaking (i.e., the process by which federal agencies formulate, amend, or repeal a regulation and the process permitting any interested party to petition for the issuance, amendment, or repeal of a regulation). Substantive regulations promulgated by an agency under the APA have the force and effect of law. FACA requires that certain kinds of groups whose advice is relied upon by the government be chartered as advisory committees, that they be constituted to provide balance, to avoid a conflict of interest, and to hold committee meetings in public with an opportunity for comment from those outside the committee. The FOIA provides the public with a statutory right to access federal agency information.

U.S. food safety programs are risk-based to ensure the public is protected from health risks of unsafe foods. Decisions within these programs are inherently science-based and involve risk analyses. Risk assessment is useful in understanding the magnitude of the problem faced, and it assists the agency in determining an appropriate risk management response.

The regulatory development process is conducted in an open and transparent manner. Regulations are developed and revised in a public process that not only allows, but encourages, participation by the regulated industry, consumers, and other stakeholders throughout the development and promulgation of a regulation. In developing new regulations and revising existing regulations, the agencies often provide the public a preliminary discussion and opportunity for comment by publishing an Advance Notice of Proposed Rulemaking (ANPR). It lays out the issues, presents the agency's suggested resolution, and solicits alternative solutions. The information received from the public is used by the agency to decide whether and how to pursue rulemaking further. All significant public comments must be addressed in the final regulation. The next steps are publication of a proposed regulation and publication of a final regulation, which is enforceable, with opportunities for public comment. The APA requires that the final regulation be justified by policy rationale, scientific bases, and legal authority.

When confronted by a particularly complex issue where advice is needed from experts who are not part of the agency, the regulatory agency may choose to hold a public meeting or convene an advisory committee meeting.

Open, public meetings, structured according to the agency's needs, bring together experts and stakeholders via an informal process. These meetings are used to receive the public's input on a specific subject area or on the agency's future programs. An advisory committee meeting is structured more formally. Public meetings and advisory committee meetings are announced in the *Federal Register* and the meetings are held in public unless an exempt issue, such as trade secrets, confidential commercial information, or personal medical information, is being discussed.

If a person or organization wishes to challenge an agency decision, the complainant may take the agency to court. Thus, even after an agency issues a final regulation which responds to all comments received, an individual or organization may still challenge the agency decision. This legal action involves the third branch of the federal government, the judicial branch. The judiciary (the federal court system) plays a critical role in the regulatory process in that it reviews an agency's action in light of the substantive law and procedural requirements. An independent judge or panel examines the whole agency record of activity detailing what the agency did and why. If the court finds that the agency did not follow its statutory mandates, fulfil the procedural requirements, or have a rational basis for its action, the judicial system can overturn the agency's action. The judicial system also serves as a forum for agency-initiated enforcement actions.

Just as it is the responsibility of the food industry to sell only safe food, it is likewise its responsibility to obey applicable laws and regulations.

## B. Risk Analysis and the U.S.'s Precautionary Approach

1. RISK ANALYSIS. Science and risk analysis are fundamental to U.S. food safety policymaking. In recent years, the federal government has focused more intently on risks associated with microbial pathogens and on reducing those risks through a comprehensive, farm-to-table approach to food safety. This policy emphasis was based on the conclusion that the risks associated with microbial pathogens are unacceptable and, to a large extent, avoidable; and that multiple interventions would be required throughout the farm-to-table chain to make real progress in reducing foodborne pathogens and the incidence of foodborne disease. This effort followed many years of concentration on managing chemical hazards from the food supply by regulation of additives, drugs, pesticides, and other chemical and physical hazards considered potentially dangerous to human health. It reflects the recognition that the

approaches to analyses and review of biological hazards and safety concerns differ from those presented by chemicals.

The President's Food Safety Initiative, announced in 1997, recognized the importance of risk assessment in achieving food safety goals. The Initiative called for all federal agencies with risk management responsibilities for food safety to establish the Interagency Risk Assessment Consortium. The Consortium is charged with advancing the science of microbial risk assessment by encouraging research to develop predictive models and other tools.

The U.S. government has completed a risk analysis on *Salmonella enteritidis* in eggs and egg products which included the first farm-to-table quantitative microbial risk assessment. It is also conducting a risk analysis for *E. coli* O157:H7 in ground beef and has entered into a cooperative agreement with Harvard University for a risk assessment of the transmission of Bovine Spongiform Encephalopathy by foods. The U.S. is also carrying out a risk analysis for *Listeria monocytogenes* in a variety of ready-to-eat foods.

Regulatory agencies also have made progress in implementing various risk management strategies. An example can be found in Hazard Analysis Critical Control Point (HACCP) regulations. Instead of including in the text of the regulation those specific steps industry must take under a HACCP system, food safety agencies provide general requirements and direct those being regulated to apply the guidelines and develop specific steps to achieve an effective HACCP program. HACCP systems are a risk management tool because they enable the user to identify hazards reasonably likely to occur and to develop a comprehensive and effective plan to prevent or control those hazards.

Performance standards for pathogen reduction and control represent another risk management tool. For example, the U.S. has in place pathogen reduction performance standards for *Salmonella* that slaughter plants and raw ground product must meet, and it also tests product to ensure that these standards are met. In the future, the government may establish performance standards for other pathogens of public health concern and define what food establishments that produce, process, or handle food must achieve.

Fair and objective regulatory decisions regarding food safety standards and requirements rely on risk analysis performed by competent authorities, qualified to make scientifically sound decisions. Risk analysis consists of risk assessment, risk management, and risk communication, which are interdependent.

*Risk Assessment*—Risk assessments are conducted in an objective manner. However, since data and scientific knowledge on any issue are never totally complete, an assessment of absolute risk is impossible. By explicitly considering uncertainties in the data and analyses, decisions can be made regarding the amount of uncertainty that is acceptable. U.S. policy decisions on procedures used for risk assessment can also ensure that risks are unlikely to be underestimated.

The first component of risk assessment, hazard identification, requires decisions on the effort expended to identify hazards. In the U.S., these are established by law and experience. Laws regarding the use of new food ingredients or pesticides require a prescribed effort to uncover any hazards before introduction into the food supply. For products already on the market, hazards may be identified by experience (e.g., emerging pathogens) that require efforts to control risk.

The second component, hazard characterization, considers data regarding the potential hazard at different exposure levels and modes, including which data are most relevant for characterizing the hazard. While human data are always most relevant, animal data are usually used to characterize a hazard. The U.S. generally relies on data from the most sensitive species to characterize the risk. Where a safety threshold cannot be assumed, the U.S. may rely on linear mathematical models that are not likely to underestimate a risk. It is important to use the most realistic data and models consistent with current scientifically sound knowledge. When information is not available that can identify which is most realistic, data or models that can be shown not to underestimate hazards are used.

The third component, exposure assessment, must differentiate between short term exposure for acute hazards and long term exposure for chronic hazards. For acute hazards, such as pathogens, data on levels of pathogens causing illness in vulnerable population groups are important. For chronic hazards, such as chemicals that may cause cumulative damage, a lifetime averaged exposure is relevant.

*Risk Management*—Risk management is exercised by highly qualified regulatory authorities with the sole objective to provide high levels of protection to the U.S. consumer. Management of risk is necessary when much, some, little, or no data are available thus requiring knowledgeable experienced experts capable of making scientifically defensible decisions in the interest of public health. Risk management principles are set by law or by the risk manager's expert judgement to reduce risk to the lowest practical, or achievable, level.

U.S. laws require that the safe use of a food additive, an animal drug, and a pesticide be established before marketing; therefore risk management decisions are based on very substantial scientific evidence. For hazardous substances that are inherent components of foods (e.g., low levels of natural toxicants produced in potatoes) or unavoidable contaminants of food (e.g., mercury in fish, aflatoxin in grains), government intervention occurs when presence of a substance reaches a level known to present significant risk. The quantity and quality of scientific evidence may vary with the type of risk management decision.

As an example of risk management, every year the U.S. federal food agencies work together to develop a comprehensive, risk-based, annual sampling plan to detect drug and chemical residues in U.S. food. Violative residue information is used as the basis for standard-setting and for enforcement and other follow-up activities.

*Risk Communication*—Routine risk communication is inherent in the transparent regulatory process which is more fully described in Part D, entitled "Transparency." Transparent standards are employed to ensure fairness to all members of the food industry while protecting public health. U.S. law requires the government to allow and consider comment on the factual basis for a decision when it establishes regulations. Anyone can comment, including persons outside the U.S. There must be a substantial basis in law and fact for every rule. Information relied on by the government is made available for anyone to review. Government scientists use public communication media to explain to the public the science behind regulations.

When there is a need for emergency risk communication, alerts are conveyed through a nationwide telecommunication system linking all levels of the food safety system with the nationwide media so all citizens are made aware of the risk, and through global information sharing mechanisms by which international organizations (WHO, FAO, Office of International Epizootics and the World Trade Organization, if appropriate), regions such as EU, and individual countries are informed immediately.

Risk communication is critical during the risk assessment and management stages. The U.S. is committed to openness and transparency of its work to protect the public from food-related health risks. For example, regulatory agencies provide public notification of recalls of food products. Information about recalls is also provided on the agency's website, as are frequent reports of regulatory and enforcement actions taken against regulated food establishments. EPA's pesticides website contains the full risk analysis for specific pesticides, and risk analyses procedures have been made available to the public for comment. Where appropriate, risk analyses processes have been modified in response to these comments.

Another example of risk management is U.S. federal agency activities on the emerging issue of resistance from the use of antimicrobials in animals. Antimicrobial risk management includes establishment of monitoring and resistance thresholds before a drug can be approved; continuous monitoring of resistance in enteric bacteria from humans and food animals; obtaining information on factors responsible for promoting resistance; and taking regulatory actions as needed, including restrictions on a drug or removing it from the market.

2. PRECAUTIONARY APPROACH. The genesis of many health, safety, and environmental laws is associated with the prevention of undesirable events and the protection of public health and the environment. Specific prevention and protection measures reflect differing provisions of law, regulation, and circumstances. However, they all are risk-based. The precautionary approach is exercised in a variety of ways.

An example of the U.S. precautionary approach to risk is the control system for ingredients in food and feed, such as the feeding prohibition of certain animal proteins to ruminants to prevent the introduction of BSE in this country. In implementing this prohibition through a regulation, the government followed existing APA procedures to explain in the *Federal Register* why it is proposed to take the action, including a description of the risk, and to evaluate the comments received from industry, academia, private citizens, and government agencies before publishing its final regulation.

Another illustrative example of the precautionary approach is the premarket approval requirements established by law for food additives, animal drugs, and pesticides. The products are not allowed on the market unless, and until, they are shown by producers to be safe to the satisfaction of the regulatory authorities. When the petition is reviewed, data are evaluated to determine exposure to the additive, including exposure to all likely impurities in the additive. The degree of testing considered necessary depends on the class of chemical and exposure. The data or the lack of data drive a decision for approval. The evaluation of all is documented. The final decision explaining the basis for all significant conclusions is published in the *Federal Register*. Persons disagreeing with the decision may file an objection with the reasons for disagreeing and request a hearing. After administrative remedies for appeal are exhausted, the government may be challenged in court on its approval or denial of a petition.

## C. Dealing with New Technologies, Products, and Responding to Problems

In achieving the nation's farm-to-table food safety objective, the federal government is only one part of the equation. Federal agencies collaborate with state and local agencies and other stakeholders to encourage food safety practices and to offer assistance to industry and consumers on practices that promote food safety.

The U.S. recognizes the regulated industry as a stakeholder and as the party principally responsible for food safety. Establishments are responsible for producing food products that meet regulatory requirements for safety. The government's role is to set appropriate standards and do what is necessary to verify that the industry is meeting those standards and other food safety requirements. Consistent with modernization of inspection systems and the farm-to-table initiatives, federal agencies use their resources as efficiently and effectively as possible to protect the public from foodborne illness. As an extension of HACCP, the U.S. is testing new meat and poultry inspection models to determine whether or not additional protections can be provided consumers through redeployment of some in-plant resources to the distribution segment of the farm-to-table chain, which includes transportation, storage, and retail sale of products.

Federal food safety agencies regularly enter into partnerships with states and others such as grower organizations and public interest groups to encourage improved production practices, to develop and foster food safety measures that can be taken on the farm and in marketing channels to decrease public health hazards in food, to develop and implement safer pest management practices, and to develop good agricultural practices to minimize pesticide residues and microbial risks.

The country's emergency response capability is sound and being enhanced continually. For example, U.S. food safety regulatory agencies participate in FoodNet, a network whose objectives are to determine the frequency and severity of foodborne diseases and the proportion of common foodborne diseases that result from eating specific foods and describe the epidemiology of new and emerging bacterial, parasitic, and viral foodborne pathogens.

Information on possible foodborne disease outbreaks from FoodNet and reports to state and local health departments are followed up by those health departments in cooperation with federal food agency authorities to determine the course and nature of the outbreak. Appropriate public advisories are issued and enforcement actions taken about the products involved as soon as possible.

In addition, a new technique has been developed using pulsed-field gel electrophoresis (PGE), which permits CDC to match distinctive patterns of pathogenic materials that cause foodborne illness. Using these "fingerprinting" techniques, the single causal factor of a foodborne illness outbreak can be traced using epidemiological investigation and PGE. This has led to intervention and, in at least one recent case, cessation of a serious foodborne illness outbreak. Both FoodNet and PulseNet are basic building blocks for the U.S. system of foodborne illness prevention.

## D. *Transparency*

Various U.S. statutes and executive orders establish procedures to ensure that regulations are developed in an open, transparent, and interactive manner and that, as appropriate, the regulatory process is similarly open to the public. Regulations and their implementation must lead to fulfillment of objectives for the public good such as protecting health, safety, and environment.

The APA specifies requirements for rulemaking (i.e., the process by which federal agencies formulate, amend, or repeal a regulation and the process permitting any interested party to petition for the issuance, amendment, or repeal of a regulation). Substantive regulations promulgated by an agency under the APA have the force and effect of law.

Under the APA, a notice of proposed rulemaking must be published in the *Federal Register*, an official daily publication which is available through subscription and through the Internet at no cost. All regulations and legal notices issued by federal agencies and the President are published in the *Federal Register*. In addition, though the Internet is not an official publication, U.S. government agencies make extensive use of it to provide information on regulatory activities and enhance the transparency of their processes.

The President issued an Executive Order to strengthen agencies' processes for promulgating regulations. Also, several states require analysis of the impacts of regulations: there are requirements to analyze the impact of the regulation on small business (the Regulatory Flexibility Act); the impact of the regulation on the environment (the National Environmental Policy Act); and the impact of any information collection requirements contained in the regulation (the Paperwork Reduction Act).

FACA requires that certain kinds of groups whose advice is relied upon by the government for establishing regulations be chartered as an advisory committee, be constituted to provide balance and to avoid conflicts of interest, and to hold its advisory meetings in public with an opportunity for comment from those outside the committee.

FOIA's purpose is to expand the areas of public access to information beyond those originally set forth in the APA. Any person residing in the United States has a right of access to a wealth of government information and records, subject only to certain limited exemptions.

To ensure the broadest possible participation by the public, agencies publish their proposals on Internet sites and call attention to the proposed or final rule through press releases. The U.S. news media and interest groups follow the *Federal Register* and agency Internet sites closely and publish information about proposed and final regulations. In addition, U.S. agencies may hold public meetings to solicit input from interested persons. Meetings often include media coverage. For example, numerous public meetings were held to solicit input on the Food Safety Strategic Plan being developed by the President's Council on Food Safety; on the draft Guide to Minimize Microbial Food Safety Hazards for Fresh Fruits and Vegetables; as part of the process to develop the Food Safety Initiative; and on bioengineered foods, among other topics.

Regulatory agencies often offer guidance on ways to achieve compliance with regulatory requirements. Such guidance may describe situations where a food could become adulterated or misbranded or may describe data that would be needed to establish safety. Although such guidance does not have the effect of law (one need not follow it to demonstrate that a food is safe and lawful, provided that all statutory and regulatory requirements are met), such advice is helpful to the food industry and to the consumer.

The Codex Alimentarius Commission (Codex) is the major international body for promoting the health and economic interests of consumers while encouraging fair international trade in food. Within the United States, Codex activities are coordinated by officials from USDA, HHS, and EPA. The U.S. Codex Office provides information via the *Federal Register* and the Internet concerning the Codex and its activities internationally and in the U.S.

## E. System Accountability

U.S. food agencies are highly accountable to government's three branches and to the people:

- U.S. food agencies are accountable to the President—the chief executive—who has constitutional responsibility to assure that laws are faithfully executed; who appoints senior officials, and whose Office of Management and Budget clears significant regulations.
- U.S. food agencies are accountable to the Congress, the legislative branch of the U.S. government, which provides the food agencies their authority and budget; whole committees hold frequent oversight hearings; and the Senate must confirm the nomination of cabinet officers and senior officials.
- U.S. food agencies are accountable to the courts, the judicial branch of the U.S. government, which review food agency regulations and enforcement actions.
- Most importantly, U.S. food agencies are accountable directly to members of the public, who regularly exercise their right to participate in the development of laws and regulations, such as commenting on proposed regulations; whose guidance is sought in frequent public meetings; and who provide strong support for food safety regulation, the nutrition label, and other regulatory initiatives.

# Index

additives, 126, 270
Africa, 188
agribusinessmen, 20
American Cancer Society, 52
American Meat Institute (AMI), 243, 258
Anderson, J. Walter, 221
Animal Liberation movement, 42, 52
antibiotics, resistance to, 91, 125, 137
Avery, Alexander, 102

*Bacillus thurigeniensis* (b.t.), 104, 119, 139
backcrossing, 104
Bailey, Britt, 134
Bailey, Ronald, 100
Baker, John, Dr., 39
Berry, Wendell, 5
Bertini, Catherine, 78
biodiversity, 130, 138, 145, 158, 172, 197–98
biopiracy, 113
biosafety, 107–8, 140
Bollgard cotton, 133–34, 184–85
Borlaug, Norman, 74, 120–21
bovine growth hormone, 80, 203
Bretton Woods Conference, 199
British Medical Association, 46
broccoli, 32–33
Brown, Lester, 33
Bt corn, 136, 140
Butz, Earl, 9, 10

cancer, 52, 124, 154
carbon dioxide, 171
Carson, Rachel, 66, 124
carrying capacity, 67–69
Cartegena Protocol, 206
cassava, 118
Centers for Disease Control (CDC), 224, 226, 253
chiliasts, 56
China, 65–66, 198
cholesterol, 52
climate, 170
Codex Alimentarius Commission, 98, 279
colonialism, 9
commercialization, 149
Cook, James, 109
cooking, 43
Cortes, 9
cotton, 133–34, 184–85
cultivars, 160–62

Davison, William L., 65
Da Vinci, Leonardo, 44
Delta Pine and Land Company, 111
DeVoto, Bernard, 8
Dolan, Dorothy, 248
Dolly (lamb), 82
DNA, 8, 92–93
dolphins, 39
Durning, Alan, 33

E. coli, 90–91, 117
E. coli O157:H7, 218, 225, 229, 251, 260
ecology, 58
efficiency, and farming, 19
eggs, 41
El Dorado, 7
energy, 61, 63
environment, 168, 173, 194
Epsy, Mike, 25–26?
ethics, situational, 58, 87
eugenics, 81
exploitation: of land, 7–10; of animals, 29–30

farms (farmers), 27, 196, 211–13
fatalism, 60
fertilizer, 35, 157, 168
Fletcher, Joseph, 64
food: as weapon, 10; security, 180
Food and Drug Administration (FDA), 96–99, 267
forests, destruction of, 3
Fox, Michael, 103
Fox, Nicols, 215

Galler, Robert and Lorie, 244
GATT, 185, 199
General Agreement on Tariffs and Trade, 185, 199
gene transfer, 92
genetic engineering, 82–84, 162, 201–4
Gandhi, Mahatma, 44
Glickman, Dan, 141
"Global Environmental Organization," 188
global warming, 170
globalization, 193–211

GM (genetically modified) food, 74–77, 116–22, 125, 148–55; as trade symbol, 186
Golden Rice, 71–72, 151, 186
Goldsmith, Oliver, 27
Green Revolution, 157, 177
Greenpeace International, 71, 102, 108, 111–12, 137
Griffin, Patricia, 239, 260

Haber Bosch process, 152
Haerlin, Benedikt, 102, 108
hamburger, 243–47
Haslberger, Alexander, 189
Hardin, Garrett, 54, 74
Hardy, Thomas, 16
Harrison, Ruth, 28
hazard analysis and critical points system (HACCP), 257–58, 273
Heersink, Mary, 215, 259
hemolytic uremic disease (HUS), 217, 232
Herbert, Richard, 223
herbicides, 160
Henry County, Kentucky, 17
Ho, Mae-Wan, 80, 102, 103, 105, 109–10, 112
Hoechst, 131
Homestead Act of 1862, 15
Horace, 66
Howard, Sir Albert, 23, 126
Human Genome Project, 89

India, 59–64, 100, 184
Industrial Revolution, 48
Integrated Crop Management (ICM), 155
International Rice Institute, 72

Jacobi, Karl, 59
Jack-in-the-Box, 245–46
James, Clive, 137
Jonas, Hans, 56

Kissinger, Henry, 190
Krebs, Sir John, 152
Kyoto Agreement, 172

labeling (GM foods), 106–7, 188
lactose, 53
Lake, Robert, 98
*Lancet* (journal), 106
Lappé, Francis Moore, 2
Lappé, Marc, 134, 156
lifeboat ethics, 54
Losey, John, 109
Lysenko, T. D., 77

McDonalds, 223–27, 230
McGloughlin, Martina, 101, 105, 110, 114
Malthus, Thomas, 56
Marx, Karl, 55
Maxted-Frost, Tanyia, 123
meat: 44, 51–53, 250; inspection, 252
Medway, Lord, 38
milk, 41, 52
Miller, Henry I., 96
modernization, 18–19
Monarch butterfly, 109, 119, 204
monocultures, 138, 161, 195
Monsanto, 112, 131, 159, 204
Moses, Edwin, 45
Multilateral Agreement on Investment (MAI), 206

nationalism, 183
nature, 76, 81, 149, 176

nitrogen, 114, 120
numbers, 56
Nurmi, Paavo, 45
nurturer, 10

organic food, 36, 107, 110, 116–22, 123, 139, 150
Organization for Economic Cooperation and Development (OECD), 89

pain (and suffering), 37
Patton, Stuart, 51
Pence, Gregory, 116
pesticides, 162–63, 174, 194, 270
philanthropy, 58, 63
Pinstrup-Andersen, Per, 115
Pioneer Corporation, 159
P. L. (Public Law) 480, 57
population growth, 55, 150, 169, 181
potato blight, 160
Prakash, C. S., 101, 110, 113
precautionary principle, 189, 272, 276
Prince Charles (England), 118, 121
Prince Phillip, 121
Princess Anne, 121
protectionism, 187
protein, 32, 46–47
Public Citizen, 102
Pusztai, Arpad, 106, 143

rapeseed, 154
Rifkin, Jeremy, 102
Right Livelihood Award, 191
Riley, Lee, 224–25, 227
risk, 271
Rockefeller Foundation, 72, 87
Rose, Murray, 45

Rotblat, Joseph, 94
rotenone, 118
Rowett Institute, 106
Roundup, 104, 132
Roundup Ready soybeans, 132–33, 145
Royal Society for Prevention of Cruelty to Animals, 38
Runge, C. Ford, 180

salmon, 38
sanctity of life, 67
Scott, Dave, 45
Seattle protests, 181, 192, 207
Sellers, James, 56
Senauer, Benjamin, 180
Shapiro, Robert, 135
Shaw, George Bernard, 30, 44
Shiva, Vandana, 101, 108, 112–15, 130
Singer, Peter, 26
Snow, C. P., 560
Sorbitol-MacConkey, 242
soybeans, 41
speciesism, 26, 29, 42
starvation, 54–79, 176, 182
substantial equivalence, 201
sugar, 188
sulphur dioxide, 176
superweeds, 139, 153, 186
Syngenta, 72

Taylor, Michael, 255–57
technology protection system (TPS), 111
technophobes, 168
terminator technology, 111, 186
Third World Network, 80, 102, 108

thrombotic thrombocytopenic purpura (TTP), 239
Tolstoy, Leo, 44
trade, 180, 193
Trade-Related Aspects of Intellectual Property, 200
tragedy of the commons, 55
transgenics, 82, 105, 136, 142, 157
transparency, 275, 278
Trewavas, Anthony J., 148, 168
triage, 64
TRIPs (Trade-Related Aspects of Intellectual Property), 200
tuna, 39

United States Department of Agriculture (USDA), 250–53
UN World Food Program, 75
Uruguay Round, 180, 183, 185
usufruct, 24

vaccines, 114
Vavilov, N. T., 77
vegans, 40, 47
Verfaillie, Hendrik, 130
viruses, 83
vitamin A, 71–72
vegetarianism, 28, 40, 43
verotoxin, 234
Vons company, 246

Walsh, Declan, 75
Walton, Bill, 45
Wambugu, Florence, 115
water usage and pollution, 34
weed, 138–39
West Wing, 74
wheat, 74
White Castle, 222

White Tower, 222
wilderness, 174
World Health Organization (WHO), 90
World Trade Organization (WTO), 98, 103, 141, 192
World Watch Institute, 33
White race, 6

# About the Editor

Gregory E. Pence has taught bioethics for more than twenty-five years in the Philosophy Department and School of Medicine at the University of Alabama at Birmingham, where he has won the Best Teacher award. He has lectured in more than two hundred universities worldwide and published in the *New York Times*, *Wall Street Journal*, and *Newsweek*. He has published *Who's Afraid of Human Cloning?*, *Re-Creating Medicine*, and *Designer Food: Mutant Harvest or Breadbasket of the World?* and edited *Flesh of My Flesh: The Ethics of Human Cloning* (all with Rowman & Littlefield). His text with McGraw-Hill, *Classic Cases in Medical Ethics*, will soon go into its fourth edition.